CAMBRIDGE COUNTY GEOGRAPHIES

SCOTLAND

General Editor: W. Murison, M.A.

LINLITHGOWSHIRE

Cambridge County Geographies

LINLITHGOWSHIRE

by

T. S. MUIR, M.A., F.R.S.G.S.
Geography Master, Royal High School, Edinburgh
Author of *Edinburgh and District*

With Maps, Diagrams, and Illustrations

Cambridge:
at the University Press
1912

CAMBRIDGE UNIVERSITY PRESS
Cambridge, New York, Melbourne, Madrid, Cape Town,
Singapore, São Paulo, Delhi, Mexico City

Cambridge University Press
The Edinburgh Building, Cambridge CB2 8RU, UK

Published in the United States of America by Cambridge University Press, New York

www.cambridge.org
Information on this title: www.cambridge.org/9781107684874

First published 1912
First paperback edition 2013

A catalogue record for this publication is available from the British Library

ISBN 978-1-107-68487-4 Paperback

CONTENTS

		PAGE
1.	County and Shire. The Origin of Linlithgowshire or West Lothian	1
2.	General Characteristics. Position and Natural Conditions	4
3.	Size. Shape. Boundaries	7
4.	Surface and General Features	8
5.	Rivers and Lakes	13
6.	Geology and Soil	18
7.	Natural History	27
8.	Along the Coast	34
9.	Coastal Gains and Losses	43
10.	Climate	45
11.	People—Race, Dialect, Population	52
12.	Agriculture—Main Cultivations, Stock, Woodlands	59
13.	Industries and Manufactures	63
14.	Mines and Minerals	67
15.	Shipping and Trade	74

PAGE

16. History of the County 79
17. Antiquities—The Roman Wall 88
18. Architecture—(*a*) Ecclesiastical 95
19. Architecture—(*b*) Military 101
20. Architecture—(*c*) Municipal and Domestic . . 105
21. Architecture—(*d*) Linlithgow Palace . . . 110
22. Communications—Roads and Railways . . . 115
23. Administration and Divisions 121
24. The Roll of Honour 125
25. The Chief Towns and Villages of Linlithgowshire . 131

ILLUSTRATIONS

	PAGE
Quadrangle, Linlithgow Palace	6
Bridge and Railway Viaduct, Linlithgow	10
Cramond Brig	14
The Avon near Wallace's Cave	16
Binny Craig	24
Linlithgow Loch	32
Queensferry, from the south	38
Kinneil House	42
Dalmeny Village	58
Bo'ness	65
Oil-works, Uphall	71
Bo'ness, from the Harbour	76
James V	82
Earl of Murray Tablet, Linlithgow	84
The "Distance Slab" from the Roman Wall	94
Dalmeny Church	96
Torphichen Preceptory	98
St Michael's Church, Linlithgow	100
Blackness Castle	102
Dundas Castle	103
Hopetoun House	107
Midhope Castle	108

PAGE

Dalmeny House 109
Linlithgow Palace . . . . *to face p.* 110
Room in which Mary Stuart was born, Linlithgow Palace . 112
South Gateway, Linlithgow Palace 114
Forth Bridge 120
Town Hall, Linlithgow 122
Mary Stuart 126
Sir James Young Simpson 129
Bathgate Academy 132
High Street, Queensferry 137
Diagrams 139

MAPS

Rainfall Map of Scotland 47
Place Names of Linlithgowshire . . . . *to face p.* 56

The portraits on pp. 82 and 129 are from photographs by Messrs T. and R. Annan; that on p. 126 from a photograph by Messrs Mansell and Co. of the picture in the National Portrait Gallery, London; the plan facing p. 110 is reproduced from Dr Ferguson's *Linlithgow Palace* by kind permission of the author; the remainder of the illustrations are from photographs by Messrs J. Valentine and Sons.

The author desires to thank Mr Jamieson, Provost, Mr Russell, Town Clerk of Linlithgow, and Mr Beveridge, Rector of Linlithgow Academy, for much useful information.

1. County and Shire. The Origin of Linlithgowshire or West Lothian.

The original meaning of the word *county* is a province governed by a count, and it is derived from Latin through French. In France a county was the portion of land which a man held by force of arms and by the consent of the people. Gradually such minor divisions of a country became permanent areas for administrative purposes. The word *shire* is of English origin, but its primitive meaning is disputed. The common view is that it means a piece of land shorn or separated from the rest by the king, who delegated a portion of his power to local governors. The best authorities, however, consider that the word means employment or care, for there is no record of any deliberate partition of the country. It is, in fact, natural to imagine that existing divisions such as parishes or thanages were combined or arranged in manageable groups, and that therefore shires came into existence by an act of union, not by an act of separation.

The origin of the officer called a shire-reeve or sheriff bears out this view. The arrangement of leaving the count supreme in his own locality did not work well in practice, and in time the king found it necessary to appoint an official to look after his private interests in these areas. This official was a kind of steward, but the duties of his modern representative, the sheriff, are now purely judicial.

The present divisions of Scotland, known as shires or counties, began to exist about the time of the Norman conquest of England, but it was not until the reign of David I that much progress was made. The English equivalent of the count was the earl, while the thane was the direct creation of the king, who thus became the fountain-head of a new aristocracy. In Scotland, under the old Celtic system, the mormaers corresponded to the French counts. The modern chief of his clan, whose influence is based entirely upon sentiment, may be regarded as a faint survival of the ancient Scottish chief, who was more of a petty king than a noble.

About half of the counties of Scotland are named after the county town. Officially, i.e. in census and other government returns, our county appears as Linlithgowshire; colloquially it is, as often as not, known as West Lothian. Much ingenuity has been spent upon the derivation of the word Linlithgow. All authorities agree as to the first syllable, which is the Welsh or Brythonic *llyn*, a pool or loch; but the rest of the word is very obscure. It is a safe inference that the whole word is Welsh or Old British. The commonly accepted derivation of *lled*, broad, and *cu*, a hollow, would carry fuller conviction were it not that, first, the hollow is not particularly broad; and secondly, the word *lled* means broader not broad. A sound principle to follow in discussing the origin of place-names is to find out the earliest spelling. In this case a document of the date 1147 has Linlitcu. This confirms the meaning already given for the third syllable; but the second must remain undecided. As might be expected, the name Linlithgow appears throughout history in many disguises. English and other foreign travellers seem to have had as much difficulty in spelling it properly, as anyone but a Scotsman has in pronouncing it. Sir William Brereton, who visited the town in 1636, writes it *Light-Gow*, which has a distant resemblance to Lithca, the present local pronunciation.

The origin of the word Lothian is less obscure. Bede has *Regio Loidis*, i.e. the same as Leeds. In the Pictish Chronicle, year 970, it appears as *Loonia*; a century later in the Old English Chronicle, it is *Lothene*. In the Welsh Pedigrees of the Saints is mentioned a prince Leudun Luydawc from the fortress of Eidyn in the North, and the district round Eidyn is named Lleuddiniawn, which, according to Professor Sir E. Anwyl, is like a Welsh derivative of Laudīnus. This prince Leudun or Lleuddin is the same whom Sir Thomas Malory calls King Lot. Hence Lothian is one of those place-names which are derived from names of persons. As a territory Lothian originally included all the country to the east of Strathclyde between the rivers Forth and Tweed.

The earliest mention of a sheriff of Linlithgow is in a charter of Malcolm IV, dated *apud castellum puellarum*—the castle of maidens or Edinburgh Castle—and in it is mentioned "Ulredus, vice-comes de Lithequ." But there is every reason to believe that the shire was created by David I, who was Prince of Lothian while his brother Alexander I was King of Scotland. The date is thus fixed as some time in the early years of the twelfth century, the golden age of Scottish history. The county stands practically the same to-day as it did eight hundred years ago. The Commissioners under the Local Government Act of 1889 made a few changes in connection with parish boundaries, but the limits of the county are well defined at least on the west and east by natural features. Sir Robert Sibbald says that under William the Lyon the area of the county was much larger than it is now, extending from Falkirk to "Penicooke," but the charters he quotes deal with ecclesiastical not civil jurisdiction.

2. General Characteristics. Position and Natural Conditions.

Linlithgowshire is a pleasingly diversified county, and is, as it were, a miniature of the Central Lowlands of Scotland. It is not a hive of industry like the Glasgow district; but the products from the shale mines have a world-wide reputation, its coal is sent to many a North Sea and Baltic port, London magazines and books are printed on its paper, shots in American mines and quarries are fired with Linlithgow safety fuses, and the bread we eat is very likely "raised" with Bo'ness yeast. The coastline is short, and the natural depth of water small;

but the docks of Bo'ness rank third in order of those on the south side of the Firth of Forth. In the main the county is agricultural and pastoral. A little more than one-half of the land is arable, about one-fourth is permanent pasture, while moorland, used for grazing, occupies only about one-twenty-fourth of the total area. Practically all the mosses and bogs which once covered a large part of the interior have been drained, and are now cultivated. The farmers of West Lothian are as enterprising and advanced as their brethren farther east.

Through the county run the principal lines of communication in Scotland, those between Edinburgh and Glasgow. Both of the routes from Edinburgh to the north traverse the shire, one of them crossing the Forth Bridge into Fife, the other touching Linlithgow on the way to Stirling. Yet in some respects the county resembles those regions on trunk routes which do not contain stopping places. The expresses thunder through the stations on their way from one city to the other. There is no town like Perth or Dundee, no junction like Carstairs, where trains simply must stop. Linlithgow is so near Edinburgh as to be under its shadow, so far from it as to derive little benefit.

Nevertheless, thanks to its mineral wealth, the county is, as we shall endeavour to show, not the least important in Scotland, while in historical interest it is surpassed by few. Linlithgow Palace as a building was in its day unrivalled in Europe for beauty of architecture and picturesqueness of situation. Within its walls have gathered in their time all of the great figures in Scottish history from James I, the poet-king, to Charles Edward, the last hope of the Stuarts. Some of the county families are among the oldest in Scotland. A Dundas of Dundas was a thane of Malcolm Canmore. Linlithgowshire gave the first martyrs to the cause of the Reformation, one of them belonging to the great family of the Hamiltons.

Quadrangle, Linlithgow Palace

It was to the Castle of Bathgate that Walter, the High Steward of Scotland, brought his bride Marjory Bruce. From that marriage descended the Stuart kings. These are but a few instances which may serve as a specimen of the very considerable importance of the county in past times.

3. Size. Shape. Boundaries.

Linlithgowshire is thirty-first in order of size among the counties of Scotland. The only counties which are smaller are Kinross and Clackmannan. The exact area is 76,861 acres, or, including water, 77,310 acres—in round numbers, 120 square miles. It forms about one two-hundred-and-fiftieth part of Scotland. The shape is that of a trapezium, the base being the Firth and estuary of the Forth, the sides are the river Almond and its tributary, the Breich Water, and the Avon with its tributary, the Drumtassie Burn. The short summit of the trapezium or truncated pyramid is an irregular line which nowhere reaches to the Forth-Clyde watershed. The coast, extending between the mouths of the Avon and the Almond, measures seventeen miles. The longest straight line that can be drawn in the county is one of twenty-two miles from north-east to south-west. But the average length is only sixteen miles, and the breadth seven miles. A walk along the boundary, including the coast, would mean a distance of about sixty miles.

On the east and south-east is Midlothian, south and south-west Lanarkshire, and west is Stirling. Beginning at the mouth of the Almond, the boundary follows that river past the village of Kirkliston up to Old Clapperton Hall, situated about a mile below Midcalder. There it quits the stream on the left bank and circling Pumpherston Oil Works returns again to the Almond about a mile below Livingston, thus giving to Edinburgh about four square miles of land which is, physically, part of Linlithgow. Proceeding up the Almond to the confluence of the Breich Water, the boundary follows that tributary nearly to its source. Striking across country in a highly irregular manner, it strides over the Almond at Harthill, and runs north-east, then north-west and west to the outlet from Hillend Reservoir. Leaving that almost immediately, it joins and follows the Drumtassie Burn to the Avon, which it clings to until the Forth is reached. The coast is remarkably uniform in outline. With the exception of Hound Point, between Cramond and Queensferry, Blackness with its ancient castle, and Bo'ness, there are no interruptions to mar the regularity of the run of the shore from east to west. At low water a wide expanse of sand is left bare except where the bottom shelves down steeply at South Queensferry. The Island of Inchgarvie is reckoned part of West Lothian.

4. Surface and General Features.

Linlithgowshire belongs wholly to the basin of the Forth, and forms part of the Central Lowlands of Scotland. The general slope is upwards from north to south, from the Forth to the southern boundary. The general character of the surface is undulating, and there is no extensive tract of level ground. Indeed, it strongly resembles that part of Midlothian which lies between the Pentland Hills and the Firth of Forth. Roads running east and west are on the whole fairly level, while those from south to north have many steep gradients. From the orographical map we learn that the general trend lines run from east to west, but there exist various departures from that rule which require explanation. The highest ground lies between Linlithgow and Bathgate, the Knock and Cairnpapple just topping 1000 feet. One cannot therefore expect to find mountainous scenery; but the ridges are here and there broken by rounded eminences, while picturesque crags lend variety to a district which would otherwise be tame and uninteresting.

Rising steeply from the shore is an embankment-like ridge which extends from Dalmeny to the Avon. Its most conspicuous summits are Mons Hill, Dundas, and Glower-o'er-em or Bonnytoun Hill, the last being 559 feet above the sea. On its summit is a tall monument erected to General A. Hope, who fell in the Indian Mutiny. From it there is a fine view of most that is characteristic of Scotland—in the distance the Highland hills; nearer, the Forth broadening out as it sweeps towards the sea; then to the south the silvery loch and the old grey town, the rich valley bounded by craggy heights; and away beyond, the pastoral slopes of the Pentlands. Westwards the eye sinks down to the valley of the Avon, and roams unchecked over the Carse of Stirling to the mountains around Loch Lomond.

South of this ridge lies the valley which is the most pronounced feature of the district, and which, beginning at Edinburgh, runs almost due west until it merges in the Carse of Stirling. It completely traverses the county, and is utilised by road, railway, and canal. At Linlithgow

Bridge and Railway Viaduct, Linlithgow

itself the valley is not more than a mile broad; and it is instructive to observe how the means of communication come close together at the town, diverging again both to the east and west. Its generally level character is apparent from the fact that the Union Canal has no locks until it approaches the junction with the Forth and Clyde Canal near Falkirk, while only one short tunnel and three aqueducts are required. The latter occur at Slateford over the Water of Leith, over the Almond about a mile from Broxburn, and over the Avon near Linlithgow; the tunnel is close to Falkirk. The word valley is used here because there is no better English word; but we must remember that no river flows or ever did flow along its course. As we shall see later, the origin must be sought for in the nature of the rocks, and in the geological history of the region.

Proceeding southwards, we mount—quickly, if in the west, if in the east, more leisurely—to a second ridge parallel to the first. Like the first, it is higher in the west. Between Linlithgow and Bathgate is a considerable extent of high ground, consisting of Bowden Hill (the supposed Mons Badonis); Cockleroy, which is not French, but probably the Gaelic *cachaileith ruaidh*, the red gate; the Torphichen Hills; and the Bathgate Hills. The average height is about 700 feet, but Knock and Cairnpapple rise above 1000 feet. Farther east the ridge becomes narrower. The Riccarton Hills are just over 800 feet; then comes Binny Craig—like the Castle Rock of Edinburgh, a "crag and tail" eminence—which may be regarded as the termination of the ridge.

Southwards again is undulating country of a moorland type, most of which has been reclaimed. The level descends to the Almond, and again to the boundary at the Breich Water, thereafter rising to the Pentland Hills, and to the low watershed between the Almond and the Calder. This part of the county is most uninteresting from the point of view of picturesqueness; but the many coalpits and brick-works with their lines of railway bear witness to its economic value. This and the Bo'ness district are the most densely populated parts of the county; and the evidence of prosperity in the flourishing towns and villages is some compensation for the commonplace landscape.

Though several mineral springs exist, Linlithgowshire has no health resorts; nor are there any fashionable watering-places. The county-town itself has a mild and salubrious climate, and seems to have been singularly immune from the plagues which in past times severely afflicted Edinburgh and even country districts in Scotland. The little hamlet of Blackness, the ancient but now decayed port of Linlithgow, presents many attractions to the lover of a quiet resting place; but the distance of four miles from the nearest railway station fortunately, perhaps, hinders its development. There are no waste lands along the coast to afford space for the golf-links, which bring wealth to Fife and to East Lothian.

5. Rivers and Lakes.

Apart from the Forth, the river system of Linlithgowshire consists of two streams with their tributaries, and a few small burns which flow directly into the Forth. The Almond, which forms in the main the eastern boundary, rises in Lanarkshire near the Cant Hills at a height of 800 feet. Flowing in a north-easterly direction, it enters West Lothian at the village of Harthill. There its course tends more to the east, and passing Whitburn and Blackburn, it receives the Breich Water, which is larger than itself, about a mile and a half above Livingston. A short distance below that hamlet, Midlothian appropriates both banks for a mile, but afterwards the Almond forms the boundary between the shires down to the Firth. The general direction of its course throughout is north-east, and its length is 24 miles. Thus the average fall is $33\frac{1}{3}$ feet per mile. The most picturesque part of the river is that which extends from the grounds of Craigiehall and Cammo estates to the mouth. From the door of the lovely little "temple" built by a former proprietor, one looks over a scene of charming beauty, handsome trees and grassy lawns which one would not expect to find in bleak Scotland. At Cramond Bridge the bed is composed of slabs of sandstone which make a fine contrast with the masses of foliage on both banks. A mile below we come to the pretty village of Cramond with its fine church and old-fashioned houses and gardens, and, on the Linlithgow side, the home park of Dalmeny. A few small fishing-boats harbour here, but navigation is difficult

Cramond Brig

even at high tide, for the channel is a devious one across the Drum Sands. A mile from the shore the river reaches the deeper waters of the Firth by the western side of Cramond Island, which can be visited dry-shod at low water.

The Avon rises on Garbethill Muir, and for two miles forms the boundary between Stirlingshire and the detached portion of Dumbartonshire. Then for half-a-mile it separates Lanarkshire from Stirlingshire. Flowing for six miles through the latter county, it touches Linlithgowshire at the confluence with the Drumtassie Burn near Avonbridge. Thence to the sea it forms the western boundary. Its total length is 18½ miles, during 10 of which it is associated with the county. Like the Almond, the general direction of its course is north-east; but near the mouth it performs a double bend, first to the north-west, then to the north-east. The cause of this is the final slopes of the ridge running from Dalmeny and terminating in Bonnytoun Hill. The height of the source is 600 feet above sea-level; and the average fall is a little more than 32 feet per mile, rather less than that of the Almond. It is joined by several tributaries, of which the most considerable is the Drumtassie Burn from Hillend Reservoir in Lanarkshire. The Brunton Burn rises on the slopes of Cockleroy, and passing near Torphichen reaches the Avon by a romantic wooded dell. At the mouth is a cavity called Wallace's Cave, where the hero is said to have taken refuge after the battle of Falkirk. Near Linlithgow the Avon receives the Loch Burn, the overflow from Linlithgow Loch.

During its course beside the county, the Almond has for the most part well-wooded banks. Its lower reaches are through country which teems with historical interest. The priory at Manuel, that is, Emanuel, now vanished, dates back to 1156, when a priory of nuns was chartered by William the Lyon. At Inveravon, the Roman Wall crossed the river on its way to Bridgeness. Near there is

The Avon near Wallace's Cave

a great bed of shells 9 feet thick and 100 yards long, which some geologists term a natural deposit, others a fine example of a "kitchen-midden," or prehistoric refuse heap. But the most interesting spot of all is Linlithgow Bridge, which must always have been an important crossing-place. Near it several battles have taken place, including that in 1526 when Lennox sought to free

James V from Arran and Angus. To this day the magistrates and Town Council of Linlithgow drive annually in procession to the bridge as forming the western march of the burgh's property.

For about a mile from its mouth the Avon meanders over a stretch of low level ground, sometimes called the Carse of Kinneil. Thence to low-water mark, it is nearly two miles of muddy sand, the haunt of sea-fowl.

To name the numerous burns flowing from the ridges of high ground into the Almond, or the Avon, or directly into the Firth, is unnecessary.

Apart from ponds, the only natural sheet of water is Linlithgow Loch. In the parish of Torphichen there once existed a shallow lake named Lochcot; but it was drained and its bed converted into arable land. Numerous reservoirs are scattered over the area; but the total amount of water-surface in the county is only 449 acres.

Linlithgow Loch is situated at the south foot of Bonnytoun Hill, at a height of 150 feet above sea-level. Its shape is an irregular oval, its greatest length—about three-quarters of a mile—being from south-west to north-east, while the broadest part, which is at the west end, measures about 500 yards. It is everywhere very shallow, the depth varying from 10 feet at the eastern end to nearly 50 feet in one place at the west. The area is 102 acres. Half-a-dozen small islands dot its surface, the homes of carefully protected swans and Canadian geese. On a promontory projecting from the middle of the south shore stands Linlithgow Palace. That and houses and gardens of the old burgh creeping down to the water's edge give character to what would otherwise be a very ordinary sheet of water. A tiny burn flows into it at the south-east corner and a rill tumbles down from Glower-o'er-em, but it is mostly fed from springs. "Linlithgow for wells" is an old proverbial expression. The bottom favours the existence of eels, which abound and which have been famous from early times. The eel-ark, a kind of wicker basket with a narrow mouth, is mentioned as far back as the days of David I. Pike and perch are numerous, but the nature of the feeding does not seem to suit trout. The outlet is at the west end, and the waters of the burn are used by two paper mills before it reaches the Avon, a short mile from its source.

6. Geology and Soil.

Rocks are classified according to their origin as Igneous, Sedimentary, and Metamorphic. When the earth was young, it consisted of a mass of molten material at a very high temperature. The lighter parts would naturally rise to the surface just as scum forms on the top of a boiling liquid. The scum is what we now know as the crust of the earth. Gradually the crust hardened until it had cooled down sufficiently to permit of the liberation of the various gases which compose the atmosphere. Further cooling condensed the water vapour which fell as rain and collected in the hollows. This water was still hot and acted powerfully upon the as yet only partially consolidated surface. The rotation of the earth set up currents in air and water. In our latitudes these currents ran, as they do to-day, from south-west to north-east. As the water moved it carried with it the sediment derived from chemical decomposition and its mechanical action served to create more sediment. Plainly the result would be long banks of débris with hollows between, running from south-west to north-east. A glance at a map of the British Isles will prove that that is the trend of the surface features in the northern halves of Great Britain and Ireland.

In time the sediment became by pressure solid rock, so that the first rocks were sedimentary or water-formed. Subsequent formations naturally conformed to existing conditions. While it is true, therefore, that everywhere in north temperate latitudes within the influence of the prevailing south-westerly winds and currents the primitive rocks are found to run in this direction, even younger rocks preserve the same course. It must be remembered, however, that the numerous disturbances which have since occurred in the crust of the earth, whereby oceans have been created and mountain ranges elevated, have seriously altered and even demolished these early conditions in many places.

During long ages rivers have carried down mud and sand from the mountains and spread them out in layers under water to become consolidated into rock. Vents have opened up in weak parts of the crust and poured forth vast masses of lava. Great movements have taken place which have brought deep-seated rocks to the surface, or even raised them up into ranges, or plateaux. Existing rocks have been so worked upon by heat or pressure as entirely to change in character. It is possible, nevertheless, to give a simple classification which will include all varieties of rocks:

I.	Igneous	*a.* Plutonic, i.e. deep-seated cooling, e.g. granite.
		b. Volcanic, i.e. cooling on or near the surface, e.g. basalt, lava.
II.	Sedimentary	*a.* Mechanical, e.g. sandstone, shale, coal.
		b. Chemical, e.g. limestone.
		c. Organic, e.g. chalk, coral.
III.	Metamorphic	*a.* Mainly by heat, e.g. marble.
		b. Mainly by pressure, e.g. slate.

The rocks with which we are chiefly concerned in Linlithgowshire are the volcanic—basalt and lava, or the various varieties of eruptive and intrusive rocks—and, most important, the limestones, sandstones, shales, and coal-measures, which make up the great Carboniferous Series.

The outstanding features in the geological history of Linlithgowshire are the deposition of the various strata belonging to the Carboniferous Epoch, the volcanic outbreaks which occurred during that epoch, the action of the great ice-sheet during the Glacial Age, and the accompanying and subsequent depressions and elevations of the land. Seams of coal and ironstone exist over the whole area from the Firth of Clyde to beyond the Firth of Forth. We have to imagine this region enjoying a moist hot-house climate, covered with strange and exotic vegetation, and alternately raised above and depressed below salt or brackish water. "Coal," says Professor Jukes-Browne, "consists of compressed vegetable matter which has undergone certain chemical changes in the course of time." The pressure was supplied by beds of sandstone and clay, or, when the water was clear, limestone, brought down by the numerous rivers, or abstracted from the salt water. Bed after bed was laid down, each succeeding layer depressing the masses below, and so preparing the way for further depositions. Thus in the course of time a very considerable thickness of rock was formed, although the water-covering was never at any one period otherwise than shallow. Fossils found in the rocks prove that the water was sometimes fresh or brackish and sometimes salt, implying a low marshy land dotted with lagoons, bordered by the sea, and traversed by the estuaries of sluggish streams. It was a time of rapid growth and equally rapid decay. Fish and marsh-loving animals abounded, but man had not yet made his appearance.

Suddenly the somewhat lethargic peace of the scene was broken. Vents opened in the ground, from which lava flowed, and scoriae and dust were shot out to lie thickly over the land. Molten matter welled up in long cracks, solidifying into rock of great hardness. Wherever the highly heated lava or basalt, trap or dolerite, came in contact with the coal, a few inches of the outside of the latter were destroyed; but the general effect, strange to say, was beneficent. Much of the impurity was driven off, as happens in the transformation of wood into charcoal, or of gas coal into coke; and a natural anthracite was formed. This coal is at its best in South Wales, whence comes the famous Welsh steam-coal, but quantities of good quality are mined in Linlithgowshire and Fife.

A long period of quiescence followed, during which the climate gradually became cooler. The ordinary denuding agencies, rain and running water, had full play. The volcanic rocks being harder offered greater resistance than the softer sandstones and shales. Hence an approximation took place to the physiography of the present day, outstanding bosses and ridges of volcanic rock with valleys between, trending from east to west. Thick forest, now of a more temperate character, covered the whole region.

From causes that are still undetermined, the temperature steadily dropped, until on the Highlands and over the Southern Uplands great ice-sheets were formed. From them mighty glaciers crawled slowly down over the Lowlands, overwhelming the tops of the highest hills, their surfaces dotted with huge boulders, by their banks lofty moraines, and pushing resistlessly before them great heaps of débris. The Forth Glacier, as it may be named, must have been at least 3000 feet thick, and by its planing work alone, it profoundly modified the face of the district. The direction of its course was first from north-west to south-east, as may easily be guessed from the fact that the main stream came from the Highlands; but afterwards, diverted by the tributary from the south, its course turned to the east. This is proved by the scratches or glacial striae found on rock-surfaces in many parts of the county.

What, then, were the main results of the glacial epoch? First, previously existing natural features were accentuated, valleys were deepened, and consequently masses of hard rocks became more prominent. Binny Craig is 718 feet above the sea, and shows a precipitous face above the hollow on the west side, which is 120 feet below the summit. It is thus an even more remarkable example of "crag and tail" than the Castle Rock of Edinburgh. Again the long valley in the centre of the county utilised by road, railway, and canal, is undoubtedly a product of the glacial epoch. Secondly, the ice-sheet covered the whole region with a layer of clay mixed with stones, from which most of the present surface soil is derived. This till or boulder-clay filled up all the river channels. In the Breich valley it has a thickness of 100 feet. Near Bo'ness borings have revealed the old valley of the Forth at a depth of more than 500 feet below the present level. Now part of that depth is accounted for by subsidence long before the glacial epoch; but it is known that before the ice-sheet came, the Almond flowed into the Forth near Dalmeny in a course 200 feet below sea-level to-day, and about a mile to the west of the existing mouth. Thirdly, while the till is mostly derived from the underlying rocks, the glacier carried on its surface boulders, which travelled for long distances. When the ice melted the boulders were left high and dry. Hence the existence of perched blocks of schist or gneiss far from their place of origin.

Binny Craig

The deposition of such a thick mantle of clay seriously affected the drainage of the country. When we know that such a vast feature as the Great Lakes of North America is consequent upon the Ice Age, we need not be surprised at transformations having occurred in Scotland. The numerous lakes which at one time dotted the surface of the Lothians, of which only one survives in Linlithgow, were hollows in the irregular surface of the till. The streams had again to set to work cutting out new channels for themselves. In most cases these coincided with the old, for the influence of the ridges of volcanic rock still persisted. But in no case have they cut so deep as they did before the glacial epoch; and here and there a stream has followed a new course. The best example is that of the Almond, whose present mouth is a mile to the east of the old one.

The surface soil is naturally derived from the underlying rock. Obviously it is thin on the hills, thick in sheltered valleys and hollows. Most of the soil in the county is of good quality, for including land that is built upon, or utilised by coal or shale mines, roads, railways, and canals, only 10,000 acres, or about one-eighth of the whole, can be described as "waste." John Penney gives 1700 acres for the area of the mosses, and these have been greatly reduced since his day (1754—1823). The prevailing soils are clay, loam, light gravel, and sand. The most fertile parts are in the central valley, and in the parish of Dalmeny. Alluvial soil exists in the "haughs" or level portions alongside the streams, as in the lower Almond and Avon.

According to Sir Archibald Geikie, the igneous rocks of our county are volcanic ash or tufa, greenstone or basalt, and trap dykes. The more prominent hills consist of intrusive rock, chiefly basalt, which is occasionally columnar in structure, as on the south side of Dundas Hill, where the columns are 60 to 70 feet in height. Cockleroy is a great mass of quartz dolerite. At Broxburn is a sill of dolerite 230 feet thick. This rock is much used for road-metal. Volcanic ash is found in various places, but there is a good exposure in St Magdalen's Quarry, near Linlithgow. The most continuous trap dyke is that which extends from Parkly Craig to the Canal Aqueduct over the Avon, a distance of 4 miles.

The Carboniferous series is as follows[1]:

I.	Coal-Measures	Upper Red Sandstone. Upper Series of Coal Seams. Lower Series of Coal Seams.
II.	Millstone Grit or Moorstone Rock.	
III.	Carboniferous Limestone	Upper Series of Limestones. Coal-Measures. Lower Series of Limestones.
IV.	Calciferous Sandstones	Shales, Cement-stones, and Sandstones, including the Upper Oil-Shale Series. Sandstones, Cornstones, merging into the Upper Old Red Sandstone.

[1] From *The Coal Fields of Scotland*, R. W. Dron, 1902.

The Linlithgowshire coalfields are two in number—that round Bo'ness, and extending for about a mile and a half inland; and the coalfield round Bathgate. Both are of the same age, but they are completely separated by the volcanic masses of the Bathgate and Torphichen Hills. Eastwards the coal-measures crop out, and their place is taken by enormous beds of igneous rock. At Bridgeness the workings extend to one-third of a mile below low-water mark. There the seams come to an end against the boulder-clay which fills the ancient bed of the Forth. The roof is composed of a singularly plastic reddish clay, which keeps the water out, and which sinks down gradually as the coal is removed.

East of the coalfields the Carboniferous Series re-appears in the famous Oil-Shale Measures of the West Calder, Pumpherston, Broxburn, Philipstoun, and Dalmeny districts. Associated with the shale are extensive beds of limestone and sandstone. The latter have been much worked, and many of the beautiful buildings in Edinburgh, including the Scott Monument, were built of this fine stone.

7. Natural History.

The British Archipelago is of continental origin, that is, it was once united to Europe. Its fauna and flora have, therefore, been derived from the neighbouring continent. The North Sea and the English Channel have several times been dry land, and across them as across a bridge must have marched many species of animals and plants. During the Glacial Epoch, however, a complete clearance took place, except in the south of England, so that the modern population dates from the close of that era, which Professor Geikie considers to have come to an end about 80,000 years ago. From a biological as well as a geological point of view, this period is regarded as quite short, and certainly not sufficient to permit of the development of any marked variations. Twice at least since the ice retreated, dry land has joined Britain to Europe; and it was chiefly by means of these bridges that immigrants came over to repeople the desolate wastes. Naturally the hardier plants and animals arrived first, and pushed close up to the line of the vanishing ice. But, as the climate became more genial, more delicate organisms followed; and there ensued a fierce conflict for supremacy, which resulted in the present more or less stable state of equilibrium.

As the land connection was not permanent, it is obvious that the number of species in this country must be less than in Europe, and that the number will diminish towards the north-west, as being farthest from the source. It should also be noted that some species had not time to cross, and others which did cross were unable to survive the altered conditions. Lastly, migratory birds are an exception to the rule, for many species visit Scotland in summer, and go no farther south.

The flora of the British Isles is classified under the four heads of (1) Alpine; (2) Sub-Alpine; (3) Lowland; (4) Maritime or Littoral. As Linlithgow has no high land the first group is unrepresented, otherwise the flora is very similar to that of the neighbouring county of Midlothian. As already noted, three-fourths of the area is cultivated, or under permanent grass, so that for wild plants we have to go to the hedgerows and roadsides, to the seashore, and to the 3000 acres or so of heath-land. At Kinneil, near Bo'ness, is the one natural copse-wood in the county, and it contains several rare plants. Of these may be mentioned *Arabis turrita* or tower wall cress, *Betonica officinalis* or wood betony, *Sium latifolium* or water parsnip, *Arum maculatum* or lords and ladies, *Geranium phaeum* or dusky crane's bill. The low shore between Bo'ness and the mouth of the Avon favours salt-marsh-loving plants, such as the Isle of Man cabbage, *Aster trifolium* or Michaelmas daisy, *Juncus maritimus* or sea rush, *Triglochin maritimum* or seaside arrowgrass. On the drier parts of the seashore from Dalmeny to Carriden grows *Cochlearia officinalis* or scurvy grass; and in summer the large golden yellow flower of the sea poppy is conspicuous. From April to September the pink or white sea-pink flourishes. Among the rarer shore plants are *Hordeum maritimum* or squirrel-tail grass, *Reseda alba* or white mignonette, *Arenaria viscosa* or sandwort. In woods and shady places the white wood anemone, yellow lesser celandine, yellow crowfoot, wild strawberry, fragrant hairy violet, are common. More rare are the dewberry, found near Bo'ness, the sweet violet, knotty crane's bill, large flowered St John's wort, and the umbelliferous jagged chickweed.

Linlithgow Loch and the smaller sheets of water,

as well as ditches and the canal, provide a habitat for fresh-water plants. Reeds and rushes line the banks, and the still patches are covered with water-lilies, both white and yellow. There are many varieties of pond-weed and duckweed, including the gibbous duckweed, while the alternate flowered milfoil, the water soldier, the flowering rush, cat's tail, unarmed hornwort, are the less common kinds occurring in suitable places throughout the county. On river banks and in bogs are the *Nasturtium sylvestre* or creeping yellow watercress, the *Anagallis tenella* or bog pimpernel, while the cultivated fields are in summer and autumn gay with scarlet poppies or yellow mustard.

Despite the fact that there is only one "natural" wood, the area planted with trees, nearly one-sixteenth of the whole, compares favourably with many other counties. The oak, birch, chestnut, beech, yew, fir, larch, are common. In Dalmeny woods the wild pear tree and the crab apple are found; but the most remarkable introduced trees are on the Hopetoun Estate, which is famed for its beech avenue, as well as for its cedars planted in 1748, sweet chestnut, silver fir, tulip trees, hemlock, spruce, plane, and deodar. The noble oaks on the west side of Linlithgow Palace are the remains of an avenue, probably planted by James I, leading up to what was then the principal entrance.

In addition to trees, several plants have been introduced at various times and in various ways. Seeds lie among the pit-props landed at Bo'ness from the Baltic, and large numbers are hidden among the ballast which is dumped down by coal-boats on obtaining a cargo. Some are "escapes" from gardens, others again are deliberately planted. For example, Patrick Murray, Baron of Livingston, made a botanic garden containing over 1000 plants. He, however, died young, and the collection was taken to Edinburgh by Sir Andrew Balfour, to form the first Botanic Garden in the city.

Out of an estimated total of some 10,000 species of animals in the Forth area, 6865 have been already recorded. As this region extends from the Highlands to the North Sea there are three subdivisions—the lower, the middle, and the upper. Linlithgowshire forms a boundary district between the lower and the middle. Altogether it would be a fair approximation to assign about 4000 species to the county. It is obviously impossible within the limits of this book to do more than glance at the most prominent features of this large number.

The simplest forms of animal life are the Protozoa, creatures consisting of a single cell. Three hundred and six species have been found in salt, brackish, and fresh water. In pools Rotifers or wheel animalcula are common and provide interesting objects for study. Other curious creatures are those which sometimes cause phosphorescence on the Forth in summer. About 50 species of sponges occur, some in the Forth, others in fresh-water, such as that of the Union Canal. Jelly-fish and sea-anemones are not plentiful, except a few common varieties. In this connection we must not forget "Grannie," the sea-anemone kept by Sir John Dalyell at Binns, which died in 1887 at the advanced age of 66. Star-fishes, sea-urchins, and sea-cucumbers number only 34 species, but amongst these is *Brissopsis lyrifera*, a heart-urchin which is rare on the east coast, although other varieties are more common. Among the Vermes, the most striking is the sea-mouse, which is a beautiful rainbow coloured creature, and whose upper side is far from resembling a worm.

Linlithgow Loch

Of Mollusca, 243 marine and 96 fresh-water species are on record, a small list compared with the Clyde area.

Nine species of Chernetidea or false scorpions occur in the Forth area, and of these eight are found in West Lothian, amongst them being the interesting *Chernes dubius*.

The nine orders of Insecta contain more species than

all the rest, amounting to 3848 for the Forth area. Beetles alone number 1300 species. Butterflies, for a reason unknown, have become in recent years much less abundant. It is possible that the growth of the coal and shale-mining industries may have had some influence.

Excluding one or two fresh-water introductions, the fishes comprise 143 species. Trout are found in the Almond, Avon, and other streams. Linlithgow Loch has been famous for its eels from the dawn of history. Constant supplies were obtained for the royal larder. In the reign of Alexander III the sheriff took toll of 800 eels per annum. The loch also contains pike and perch, while dace were introduced and are flourishing. Marine species are naturally not so abundant in the estuary as they are farther to the east. Of the rarer kinds, the power cod is occasionally met with, and a sting ray was once caught above Queensferry. At certain seasons cod are plentiful on the mud banks fringing the coast. Salmon are fished for with nets; and in 1893 a salmon net at Bo'ness contained a sword-fish 8 feet 2 inches in length, the "sword" being 2 feet 5 inches long.

In the Forth area 253 species of birds have been recorded—divided into land birds, shore birds, and sea birds. Sea birds are not numerous in Linlithgowshire. In the Statistical Account are mentioned pheasant, partridge, woodcock, snipe, wild-duck, cuckoo, blackbird, wood-pigeon, missel-thrush, crested lapwing, grey plover, and waders. Large numbers of starlings nest in some years in a dwarf fir wood on Cramond Island. In Linlithgow Loch are swans, while Canadian geese have been introduced, and are breeding. The little auk is sometimes picked up on the coast after a gale. The tufted duck's nest has been found on a pond near Kirkliston.

Owing to the large area under cultivation mammals cannot be expected to be abundant. The Norway rat has exterminated the native rat. The wild cat, marten, and polecat are now extinct. Weasels and stoats are numerous, as well as hares and rabbits. The fox is strictly preserved for purposes of sport. The Linlithgow and Stirlingshire Hunt, originated about 1775, still flourishes, though it covers a smaller area than formerly. A few hundred fallow deer adorn Hopetoun Park. The badger was introduced by Lord Rosebery into Dalmeny Estate in 1889, and has spread westward. Seals and porpoises are occasionally captured, and a whale is sometimes cast up on the shore. A mammoth's tusk was dug up near the Almond in 1820 during excavation for the canal; and we know that these large animals as well as reindeer, and woolly rhinoceroses, roamed over the pre-glacial and post-glacial Lothians.

8. Along the Coast.

A short walk of seventeen miles will enable a traveller to see the whole coast-line of Linlithgow. Only two towns—South Queensferry and Bo'ness—break the shore, so that the scenery is mainly rural. The tourist, unless he be specially interested in natural history or geology, would do well to choose a period about high water, for the broad, flat, uncovered beach presents an expanse of mud and sand which is somewhat dreary from the artistic point of view. The best time of year is from May to the middle of October, for then the wealth of foliage, which is the special attraction, is at its best.

Beginning at the Almond we cross by the "free ferry," which is maintained by Lord Rosebery. The river here is tidal, and the average width is about 15 yards. The deepest water is on the Linlithgow side where the bank is steep. Looking Firthwards we see the winding channel of the stream marked by posts out to Cramond Island, a mile away across the sands. A well-kept path leads close to the beach, past the Hunter's Craig or Eagle Rock, and Snab Point, which are projecting masses of trap rock. From the Almond to Hound Point, a distance of two and a quarter miles in a straight line, the foreshore is a broad stretch of sand, called the Drum Sands. A few small rocks, the Buchans, alone break the monotony of the surface, which, at low water, is alive with shore birds.

From Snab Point the path descends a few feet, and passing a wall of rhododendrons, which conceals the home farm, enters the park. At this point we cross a small burn, which very possibly marks the pre-glacial course of the Almond. Here there is a wide expanse of level ground, and beyond is an abrupt rise marking the position of the ancient 25 foot sea-beach. On the edge of this is built Dalmeny House, one of the residences of the Earl of Rosebery. The word Dalmeny has nothing to do with "dal" or "dale." The old spelling is Dumanie or

Dunmanyn, meaning the hill of Manau, the name of an ancient Pictish tribe and district. Dalmeny House is a comparatively modern building, but close by is the site of Barnbougle Castle (Brythonic for shepherd's tree), which is probably the site of the residence of the Moubrays, the owners of the land during the thirteenth century.

Passing between Dalmeny House and Barnbougle, we reach Hound Point, a bold headland formed by a spur of Mons Hill. Rising within a quarter of a mile of the Firth to a height of over 300 feet, the hill is a mass of intrusive volcanic rock which has been exposed by the denudation of the softer material around and above it. From the summit the eye glances over an extensive and beautiful view. Sir Walter Scott's classic description of the Firth with its "emeralds chased in gold" at once occurs to the mind. But the admixture of woodlands and cultivated fields, craggy heights and rounded eminences, islands, winding shores and broad water makes up a scene which is admirably typical of a Scottish landscape at its best. Nor can we leave out the hand of man, for there, less than a mile away, is the Forth Bridge, even now, twenty years after its opening, one of the wonders of engineering, and, a short distance beyond it, Rosyth, the Scottish Portsmouth. Descending, we leave Dalmeny Estate at the Long Craig Gate, and, in a few hundred yards join the Edinburgh Road at the foot of the "Hawes Brae" just underneath the bridge.

Here is the famous Hawes Inn, where David Balfour was kidnapped at the instance of his rascally uncle. Here also "the Antiquary" halted on his way to Arbroath.

But merely to enumerate the characters in fiction and in history who have used the celebrated ferry, would occupy a great deal of space. One we must not omit, Queen Margaret, wife of Malcolm Canmore, after whom was named this "Passagium Reginae." The intrusive dolerite which forms Mons Hill and Hound Point runs across the Firth by Inchgarvie and reappears in the North Ferry Hills. These project far into the water, causing a narrowing of the channel which for centuries made Queensferry the only passage from Edinburgh to Dunfermline, except for the circuitous route round by Stirling. Now it is utilised by the Forth Bridge, and forms an important link in the railway from London to Aberdeen. At this narrow part we note that the low sands are absent, not to reappear until the shores begin to recede again. The ferry itself is still maintained, and a small steamer carries across at regular intervals a cargo of motor cars, tradesmen's carts, cyclists, and foot passengers.

Half a mile above the bridge lies the old royal burgh of Queensferry, nestling at the foot of an escarpment which represents an old sea cliff. A mere village in size, it is a busy place when "the fleet" is at anchor below the bridge. Then there is a constant passing to and fro of steam pinnace or long boat, bringing batches of tars on leave, or taking the mails and loads of provisions. The main, and, indeed, the only street is narrow and tortuous, for Queensferry is a terminus, the three main roads from inland stop there. Yet it has piers and a tidal harbour, whence a single line of railway climbs up to the "trunk" route, high above.

Queensferry, from the south

About half a mile inland is the little village of Dalmeny with its ancient and beautiful parish church. Close by are the Oil Works which tap a continuation of the Broxburn shale field. A short distance west is the estate of Dundas. The Dundas family is probably the oldest in the county, as it was noble in the days of Malcolm Canmore. The family possessions include the Island of Inchgarvie, which was granted by James IV. In the park is the Lily Loch Reservoir, which provides Queensferry with part of its water supply.

From Queensferry a rough road leads westwards past the harbour, across the single line of railway, and joins a better road coming down from Kirkliston. Onwards to the Avon the coast assumes its former character of wide flats uncovered at low water. One part projects well into the estuary, and is named Hopetoun Bank, after the estate which we now enter. At the quaint hamlet with the quaint name of Society, the road turns sharply inland, and eventually joins the main road from Edinburgh to Linlithgow. From it branches lead Forthwards to Abercorn and Blackness; but it is not till Bridgeness is reached that it is possible to drive or cycle beside the Firth. For some distance, however, the public road passes through Hopetoun Park, close to and in full view of Hopetoun House. Here is the magnificent beech avenue referred to elsewhere, while behind the house is a large garden laid out in a style to suit the classical character of the building.

Returning to the beach, we come presently to the Haugh Burn or Cornar, from which Abercorn takes its name. Rising 88 feet above the present shore-line is an interesting exposure of an ancient sea-beach, from which bones of whales have been taken. Beside the burn stands the parish church, which relies for its congregation more upon the oil-workers of Philipstoun than upon the small population immediately around. About a mile inland is the estate and tower of the Binns, where lived the notorious Dalyell, the scourge of the Covenanters.

We note from the map that the Forth is gradually shallowing, and that the shipping keeps to the deeper waters of 50 feet and more near the Fifeshire side. This, as well as difficulty of land access, accounts for the decline of Blackness, once the third port in Scotland. One might travel far, however, without coming to a more picturesque spot. On the point of a slender promontory stand the tall gaunt buildings surrounded by a high wall as grey as themselves. According to local tradition the Castle is one of the fortresses which the Act of Union stipulates shall be kept in constant repair. We have been unable to find in that Act, although it is mentioned in others, any reference whatever to this or to any other Scottish fortress. Yet the fact remains that the Castle is in a better state now than it often was during its long and eventful history. It is garrisoned by a lieutenant, a sergeant and 25 men and used as a powder magazine. A few men are also employed in the manufacture of fuses.

The little village itself is 250 yards away, at the head of the bay. It is in the form of a square, and the principal building is the Guildry House, a relic of the domination of Linlithgow, which was as complete as was once that of Edinburgh over Leith. To this day the magistrates, council, and faithful burgesses of King Robert the Second's "dear burgh" drive in state once a year in June down to Blackness. On a mound near the Castle the Provost, emerging from a bower of leafy branches, supposed to represent an ancient chapel of St Ninian, proclaims the election of Mr So-and-So as Baron Bailie, to look after the interests of Linlithgow in its ancient port for the ensuing twelve months. All present take a very proper pride in adhering to the time-honoured details of the picturesque ritual of the ceremony.

From Blackness a pretty foot-path takes us in two miles to Carriden House and Bridgeness. Close by the 100 foot beach is again well marked. The platform is cut out of the boulder clay, which till recently supplied excellent material for bricks and tiles. Just below Carriden House is a market garden, the outer wall of which is said to be the actual wall of the birthplace of Colonel James Gardiner, who fell at the battle of Prestonpans. Crossing a meadow, we reach the public road at Bridgeness, supposed to mark the end of the Roman Wall from the Clyde to the Forth. About a mile inland is Walton and three miles west of it is Kinneil, Gaelic for wall's end.

For nearly two miles from Bridgeness we walk between potteries and saw-mills, catching glimpses of the docks with piles of pit-props on the quays. The old town of Bo'ness is, one would think, the most inconvenient seaport in the world. If its docks are spacious, its streets are narrow and exceedingly inconvenient. Then the single line of railway meandering upwards to the main line at Manuel, and continuing to Slamannan and Airdrie with a branch to Bathgate, must lead to long delays, and hamper the growth of the port. There is no doubt, however, of the prosperity of the town.

A mile west of Bo'ness we pass beneath Kinneil

Kinneil House

House, the old family seat of the Hamiltons. It was in the dining-room of this house that James Watt first succeeded in getting a model steam engine to work. Here also lived Dr John Roebuck, the founder of the Carron Ironworks. From Kinneil to the boundary is a short mile, beside the muddy sands across which the Avon finds its way to the Forth.

The evidence of various upheavals of the land, though not continuous, is very clear in several parts of the county. All along the shore, the inner edge of the highest terrace clings closely to the 100 foot contour line. It is noticeable as far up the Almond as Kirkliston, and up the Avon to Linlithgow Bridge. At Abercorn the height is about 88 feet. In Bo'ness may be observed a platform 125 feet in height, close by an old beach at about 100 feet, while the town itself is built on the terrace of the 25 foot beach, which is conspicuous in many parts of Scotland and on which many coast towns are situated.

9. Coastal Gains and Losses.

As regards Scotland, neither elevation nor subsidence of the land is now taking place, or has taken place within historic times. The elevation, marked by the 25 foot beach, occurred after the arrival of Neolithic man. The subsequent period has been one of quiescence, so that no gains or losses are being caused by crustal movements.

The amount of coast erosion depends largely upon the nature of the shore. Where cliffs descend steeply into the sea with no marginal region of shingle, the waves have little power, unless the rock is one that naturally tends to split or is seamed with joints. Water alone has but small effects. It is where the waves are able to fling pebbles and boulders at the cliffs that the greatest destruction is observed.

The shore of the county is composed of carboniferous rocks and of intrusive volcanic masses. It is a tendency of the latter to form headlands while bays are eaten into the former. Hound Point is a conspicuous example of an igneous promontory.

Along the coast of Linlithgowshire no erosion is proceeding at the present time. Barnbougle Castle, which is built on a point near Dalmeny, is protected by a sea-wall; but there is no evidence that the waves are really at all destructive. Farther east in Midlothian, beyond Cramond, the sea is slightly more active, but not to any serious extent.

A glance at the Ordnance Survey map will show that accretion is considerably more prominent. The Drum Sands obviously consist of material brought down by the Almond. The tide retreats so far that one can reach Cramond Island on foot at low water. Clearly it is only a matter of time before a considerable area is added to the land. But it is farther west at the mouth of the Avon that the largest gains have been made. The flood-plain of the Avon is several square miles in area, about half of which is at present under water at high tide. Much of the alluvium is undoubtedly also derived from the river Forth. West of Bo'ness is the Carse of Kinneil, where reclamation has been carried on by the neighbouring landowners—the Duke of Hamilton and Mr Cadell—since 1854. The total amount reclaimed is 157 acres, divided among three farms. One of them has the suggestive name of North Hainings.

A scheme has been proposed for the building of a long embankment between Bo'ness and Grangemouth by which 2000 acres would be rescued from the Firth. The cost would be considerable, but the land would be available for agricultural purposes and for shipbuilding yards, as has been the case with similar areas at Bo'ness and Grangemouth.

The bed of the sea slopes very uniformly from Linlithgowshire across to the deep channel near the coast of Fife. Even at Queensferry, where the Firth narrows, the south side descends more gently than the north, which plunges very quickly to the depth of more than 200 feet. The only prominent sand-bank is the Hopetoun Bank, which projects into the Firth near the hamlet of Society. North of Barnbougle are the Buchans, a collection of rocks covered at high water. Opposite the mouth of the Avon, but still close to the Fife coast, is the end of the 50 foot deep channel. A mile or so farther up the depth becomes less than 25 feet. Without much dredging, then, no great sea-port can ever come into existence above Grangemouth.

The Forth Bridge is, of course, well lighted, and the two deeper channels are shown by lighthouses on each end of Inchgarvie. Pier lights are placed on the piers at South Queensferry, Blackness, Grangepans, and Bo'ness. There is a lighthouse to show the tidal harbour of Port Edgar, a little to the west of Queensferry.

10. Climate.

North-western Europe is peculiarly favoured in the matter of climate as compared with other regions in corresponding latitudes. The warm air rising from the

Equator proceeding northwards becomes cooled and descends about latitude 35° N. As it has come from a region of faster to one of slower rotation, it gains upon the earth; and instead of being a south wind, it swings round to the south-west. But for the influence of the formation of the land it would ultimately blow directly from the west, as it does in latitudes south of the Equator. It is the southerly element in the wind that causes all the difference. Not only does it bring the warm air from the mid-Atlantic, but it drives before it a surface current of warm water.

Proximity to the ocean is always an advantage, as is seen in the milder climate of Nova Scotia compared with the inland country due west; but when to that proximity are added warm currents of air and water, then the advantage is, in a sense, out of all proportion. Suppose the earth to rotate in the other direction, one result would be that the higher parts of Scotland would become permanently ice-covered like the interior of Greenland.

Not only do we gain warmth, but we have also abundance of moisture; in some cases, considering the nature of the soil, a superabundance. As the highest land is close to the west coast, precipitation is greatest there, and it diminishes with fair regularity towards the east. A rainfall map constructed on the basis of using deepening shades or colours to denote heavier precipitation would correspond very closely with an orographical map, where deeper shading or colouring denotes increasing height of land. A large number of rainfall stations are maintained throughout Scotland, and their returns make

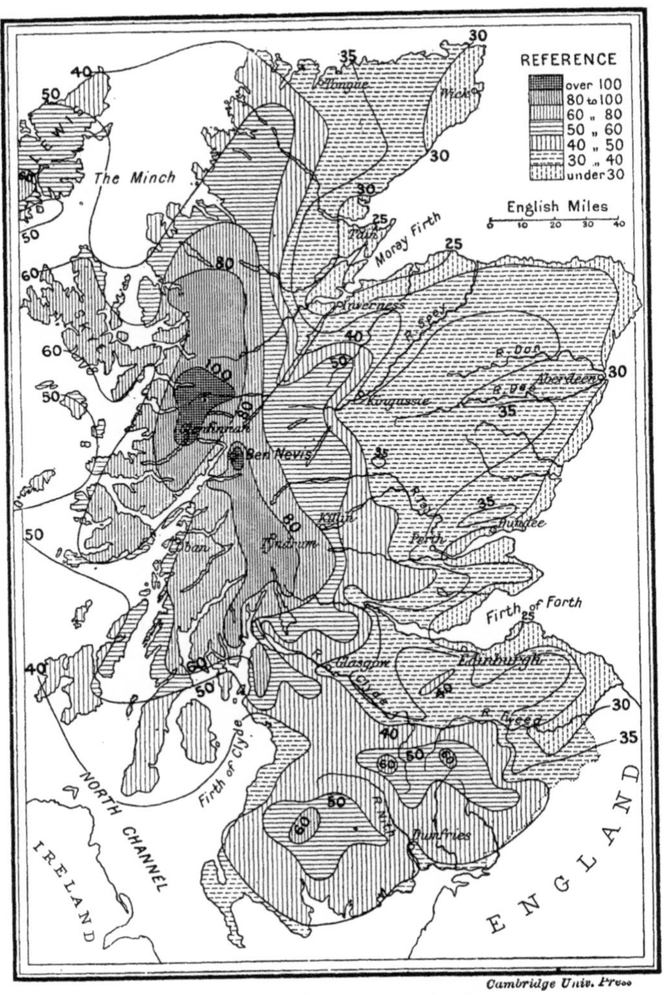

Rainfall Map of Scotland

(*By Andrew Watt, M.A.*)

possible the extremely definite map which appears on page 47.

On the average our country is not greatly favoured by sunshine. Even in summer with its long days, the amount of sunshine is smaller than in the Channel Islands, where the possible quantity is less. And in winter the deficiency is much greater than can be accounted for by merely astronomical reasons. The cause is of course the mountainous character of the surface, which encourages the formation of clouds.

A study of an isothermal map of Scotland reveals some interesting facts. In winter there is a general equality of temperature all down the west coast. In other words, Cape Wrath is, in January, as warm as the Mull of Galloway. Farther east is an approach to continental conditions; and we find small areas, for example, round Aberdeen, where the temperature is less than 38° Fahr. The tendency is, therefore, for temperature to decrease from west to east.

In summer the influence of the land becomes stronger; and the isotherms are roughly parallel to the lines of latitude. Temperature decreases from south to north. At all seasons, however, the ocean must be reckoned with. In winter it warms, in summer it cools, the land exposed to it. There is still another consideration. Isothermal maps are constructed on the ideal basis of imagining that the land is all at sea-level. An average allowance of one degree for every 280 feet must be made. For example, the summit of Ben Nevis is about 16° colder than is shown on the map. Under winter conditions, when snow covers the hills and not the lowlands, this difference becomes even greater.

With the exception of rainfall, few statistics are available for an account of the climate of Linlithgowshire; but there is no reason to believe that the conditions are vastly different from those of its neighbour, Midlothian. Reasoning from general principles, we should say that West Lothian is colder in winter and warmer in summer, distance from the sea being the factor in both cases. The rainfall we should expect to be about half-way between those of Edinburgh and Glasgow. Such figures as we possess bear out those statements.

The only temperature records are from Linlithgow Academy, where a complete series of observations has been made since 1906. The station is 170 feet above sea-level, so that, for purposes of strict comparison, about ·6 of a degree would require to be added; but as we are not constructing an isothermal map, we shall take the actual readings. Over the five years, 1906–1910, the average annual temperature is 46°·8, or practically the same as the hundred years' monthly mean at Edinburgh. The coldest month is January, but December and February are not far behind. If we include 1911, the average summer temperature is just 58°, which is higher than the Edinburgh average. The warmest month is July, but August runs it very close. The absolute maximum of 90° was reached in September 1906, and in July 1911. The absolute minimum of 6° occurred in January 1910. This gives an extreme range of 84°, while the average annual range is 20°.

It is more than probable that the Linlithgow temperatures are higher than those farther inland, not only on account of its nearness to sea-level but because of its exceptionally favourable situation. The "hollow" is well sheltered on all sides except the west, the direction from which the mild winds come. It has been noted by inhabitants that snow-storms which are especially severe in other and quite neighbouring parts of the country, are but little felt at Linlithgow, and indeed all along the peculiar valley in which it stands. Further, the meteorological station at the Academy is placed on a thick sand-bed, which is warmer than the clay predominating elsewhere.

The prevailing wind in the county, as over all Scotland, is from the west. Second to it are those from the easterly quadrant. As the line of the ridges runs from east to west, northerly and southerly winds will be deflected. The spring east winds are less severe than in Edinburgh and Haddingtonshire, while the south-easterly gales which embitter the lives of the inhabitants of the East Neuk of Fife are comparatively innocuous. In the month of May when the biting "haar" envelops Edinburgh as in a chill winding sheet, the cyclist fleeing westwards often finds himself in brilliant sunshine on reaching Linlithgowshire.

One would expect the rainfall to increase from the coast inland, and also towards the south-west, where lies the watershed between Forth and Clyde. This *a priori* statement is in the main borne out by the facts. Out of the half-dozen stations well distributed over the county, that at Polkemmet near Whitburn shows the heaviest rainfall, amounting to 44 inches on the average of nineteen years. Polkemmet is 600 feet above sea-level, and close to the border. Bathgate, 100 feet lower, and four miles to the north-east, has 40 inches. About five miles north of Bathgate is Linlithgow, with an elevation of 170 feet and a precipitation of 35 inches. Bo'ness, less than three miles northwards, has 27½ inches. Uphall, though 380 feet above the sea, and five miles inland, has only about 32 inches or less than Linlithgow. The explanations seem to be that Uphall is surrounded by higher ground on all sides except the east, and that it is the most easterly of the stations, thereby approximating to conditions in the lower parts of Midlothian. The average annual rainfall at Edinburgh is about 26 inches. Our figures clearly prove that precipitation increases as we go westwards.

The driest year since 1890 was 1902, when the rainfall nowhere touched 30 inches; and the wettest 1903, when Polkemmet recorded 55½, and even Bo'ness nearly 39 inches. The spring and the summer of 1911 will long be remembered for their exceptional drought and heat. Even by August anxiety was felt for the water-supply, and by September all those who were able had left the county for more favoured regions. The reason is that no part of the county gets water from among hills where a reasonable amount of rainfall is practically certain. Linlithgowshire is dotted all over with reservoirs; but the highest of any size are Cobbinshaw about 900 feet and Corston about 800 feet, both in

Midlothian. None of them is in a region where drought is never experienced; and all have been several times reduced to very small dimensions.

Taken as a whole, however, the climate of the county approximates to the happy medium for Scotland. It is favourable to agriculture, though rather moist for wheat, which is also not so well suited for the rather heavy cold soil as oats or barley.

11. People—Race, Dialect, Population.

It is a matter of dispute whether Palaeolithic man penetrated to Scotland. There is no doubt, however, that, as the Glacial Age came to a close, Neolithic man crept northwards on the tracks of the reindeer, and established himself all over the lowlands. These primitive people were small, dark-haired and long-skulled. They are generally named Iberians or Silurians, and their present-day representatives are the Basques, in the south-eastern corner of the Bay of Biscay. No absolutely indisputable fragments of the Iberian language remain in Scotland; but there are some place-names which are so obscure that authorities hopelessly disagree as to their derivation. It is an attractive theory that such names are survivals from the Neolithic Age; but attempts to connect them with modern Basque have proved fruitless.

After the Iberians, came the great Celtic invasion. Linlithgowshire being an open county with no secluded places for retreat, it is most likely that the original inhabitants were completely driven out. Hence the modern native of West Lothian must trace his connection with Neolithic man through the Celts. The identity of the Picts is the subject of acute controversy on account of the scantiness of the material available. Some say that their language was Gaelic or a dialect closely akin to it; but the balance of opinion leans to the theory that the Picts were Welsh or Old Britons. It is certainly true that the Gaelic-speaking missionaries from Iona usually employed interpreters when addressing Pictish audiences. Now it was in our county that the Gael and the Britons or Welsh, those two branches of the race which persist to this day, came into violent conflict with each other. One after the other must have prevailed, for Linlithgow is Welsh[1] and Torphichen is Gaelic. The ultimate success of the British kingdom of Strathclyde may be inferred from the fact that the northern boundary of Lanarkshire overlaps on the Linlithgow side of the watershed.

It does not appear that the Romans have left any traces of their presence except the remains of the *vallum* and road. Their occupation of the district was so short and so intermittent that they had no permanent influence on people or language.

After the Romans came the Angles who spread northwards and westwards to the Avon in the fifth and sixth

[1] It is possible that all of the apparently Welsh names in the county are of Pictish origin, for they are confined to the district comprised by the old Pictish kingdom of Manau, while none occur near the border of the old British kingdom of Strathclyde.

centuries. They did not gain possession of West Lothian without a struggle, and their tenure of it was never very secure. Till about the eighth century it was held by a division of the Picts, whose special designation survives in the second half of Clackmannan, Slamannan, and, according to good authority, Dalmeny. During the ninth, tenth, and eleventh centuries Norse descents were numerous; and the Danes from Ireland after landing on the Clyde sometimes pushed on by land as far as Fife. There is no record, however, of any settlements in West Lothian; and, with one exception, there are no place-names of indisputable Scandinavian origin. Bo'ness, Blackness, and Hawes are English, but Bridgeness is doubtful. There is no stream and therefore no bridge. The old spelling is Brigness, and a possible derivation is the Norse *brigg*, a landing-place, and *ness*, a cape. As far as meaning goes, that makes sense, for it agrees with the geographical conditions; but on the other hand it is unlikely that the inhabitants would name a spot in a foreign tongue.

Returning to the Angles, we note that the proportion of English names shows a marked decrease as one proceeds from East to West Lothian. And though the district was for long by the chroniclers called Saxonia, the region between the Esk and the Avon was surrendered by the Angles to the Scots in 960; while in 1018 by the great victory of Carham, Malcolm gained the whole province. Nevertheless the Angles conquered in the end, for they imposed their language on the Scots.

The Normans never came in great numbers to

Scotland, and left no mark on the race, and little on the language, though their feudal system was introduced into Lothian by David I. Such names as Champfleurie and Champany are due to the French influence during the period of the Franco-Scottish alliance (1295–1560).

The study of place-names, while not strictly a part of geography, requires a knowledge of geographical conditions. A theorist often evolves a beautiful derivation which a visit to the ground shows to be totally inapplicable. For original work in this field it is essential to have a command of the various languages, or to have access to a recognised authority[1]. It is impossible within our limits to give a complete list of the place-names in the county. That would require a treatise to itself. But with the help of the general principles here stated it will be easy to make classified lists.

The Iberians and the Romans have left no place-names. The Norse only one, Bridgeness, and that is doubtful. Two or three may be ascribed to late French influence. Hence practically all of the names may be divided into three classes—

I. Old British, Pictish, or Strathclyde Welsh, i.e. Brythonic or Cymric.

II. Gaelic.

III. English.

[1] The writer desires to offer grateful acknowledgments to Dr Watson, Rector of the Royal High School, Edinburgh, for invaluable aid on this point.

I. Distinctively British elements are *tre*, meaning house, and *aber*, meaning mouth or confluence. They are represented in the county by the names Ochiltree and Abercorn. Other Brythonic names are Calder, Almond, Pardovan, Barnbougle.

II. Gaelic was introduced subsequent upon the conquest of the district by the Gael, and the disappearance of the Picts of Manau. Certain prefixes are characteristic of Gaelic, such as *auch* field, *bal* stead, *drum* ridge, and *kin* head. Examples are Auchinbard, "field of the enclosures"; Ballencrieff, "tree-stead"; Drumduff, "black ridge"; and Kinneil, "wall's end[1]." Other Gaelic names are Avon, "*the* river"; Binny, "place of the horn" or "pointed place"; Craigengall, "rock of the foreigners."

III. English names, on the other hand, are best classified according to their suffixes. The most instructive are *ton* homestead, *law* hill, and *ing* son of. These English names form an important element, and many of them are of old standing. Some are Kirkliston, Broadlaw, and Bonnytoun. The last name is spelt in a charter of Robert I Bondington, which effectually disposes of the fanciful theory that it had any connection with the French *bon*. Other examples of place-names will be found throughout the text. Any doubtful derivations are specifically marked.

[1] The original Pictish or British form of the Gaelic Kinneil or Cenail was Peanfahel, which occurs in Bede as Peneltun. The hard C or K of Gaelic corresponds to P in British.

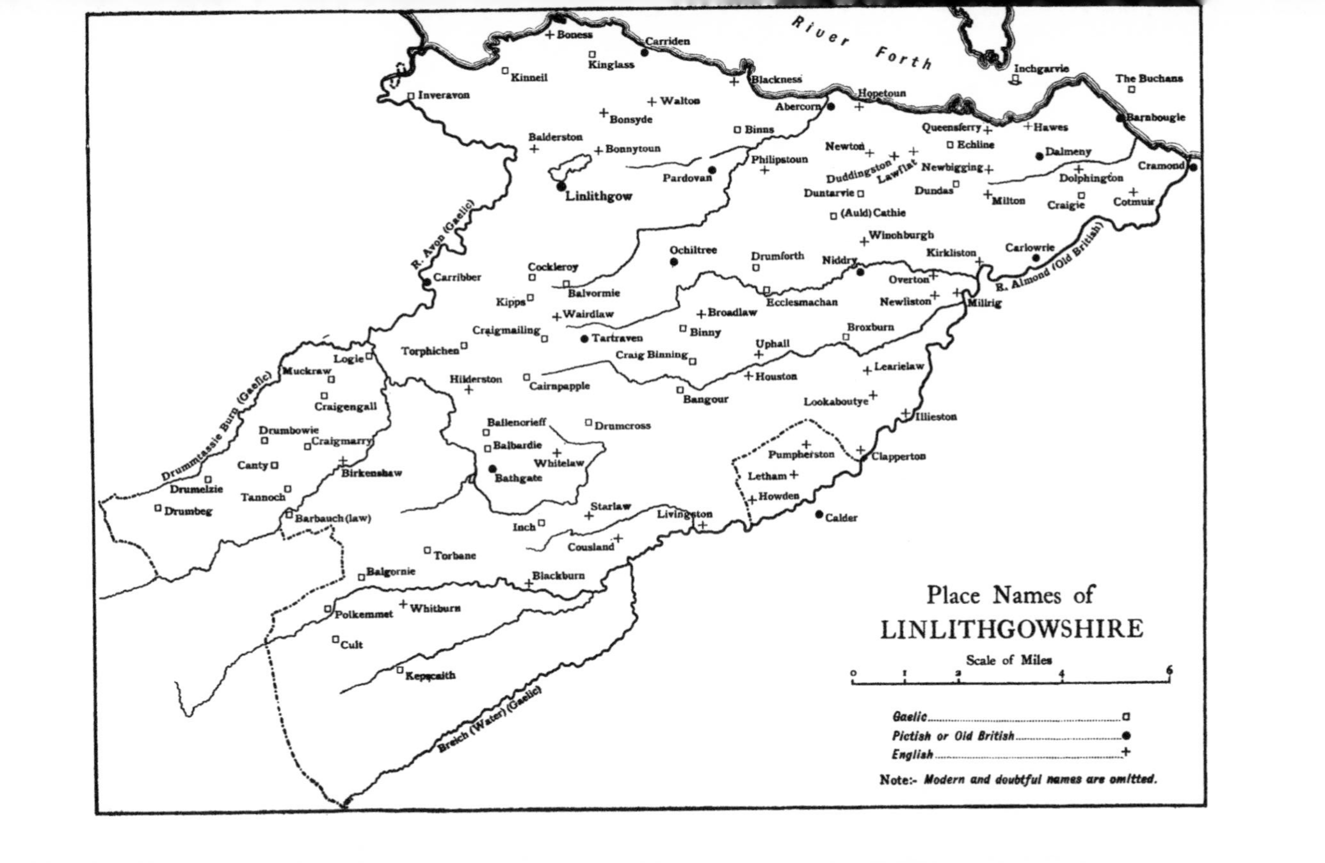

Place Names of
LINLITHGOWSHIRE
Scale of Miles
0 1 2 4 6
Gaelic
Pictish or Old British
English
Note:- Modern and doubtful names are omitted.
River Forth
Boness
Carriden
Kinglass
Kinneil
Blackness
Inveravon
Walton
Bonsyde
Balderston
Bonnytoun
Binns
Abercorn
Hopetoun
Inchgarvie
The Buchans
Barnbougie
Queensferry
Hawes
Echline
Dalmeny
Newton
Duddingston
Lawflat
Newbigging
Dolphington
Cramond
Linlithgow
Pardovan
Philipstoun
Duntarvie
Dundas
Milton
Craigie
Cotmuir
(Auld) Cathie
Winchburgh
R. Avon (Gaelic)
Ochiltree
Drumforth
Niddry
Kirkliston
Carlowrie
R. Almond (Old British)
Cockleroy
Carribber
Balvormie
Overton
Kipps
Ecclesmachan
Newliston
Millrig
Wairdlaw
Broadlaw
Craigmailing
Tartraven
Binny
Broxburn
Torphichen
Uphall
Craig Binning
Logie
Muckraw
Learielaw
Hilderston
Cairnpapple
Houston
Bangour
Craigengall
Lookaboutye
Illieston
Drumbowie
Craigmarry
Ballencrieff
Drumcross
Balbardie
Pumpherston
Clapperton
Canty
Birkenshaw
Whitelaw
Drummtassie Burn (Gaelic)
Bathgate
Letham
Drumelzie
Tannoch
Howden
Drumbeg
Starlaw
Barbauch (law)
Livingston
Calder
Inch
Cousland
Torbane
Balgornie
Blackburn
Polkemmet
Whitburn
Cult
Kepscaith
Breich (Water) (Gaelic)

The present-day inhabitants of Linlithgowshire are of the mixed race known as the Lowland Scots, the English element predominating; and the language now spoken is English. "Braid Scots" is rapidly dying out. The few old words that still survive are mostly of French origin, such as *douce*, *ashet* (for plate), *dour*. Such differences as exist are matters of pronunciation, intonation, and spelling, rather than of separate derivation. Vowels are broader and purer than in English over the border. There is no tendency towards converting a pure vowel into a diphthong as is found south of the Humber. In teaching French to Scottish children it is not necessary to tell them that it is incorrect to say *baocow* for *beaucoup*. The letter *r* gets its full value, and a little more; but *t* is often slurred or even omitted in a most objectionable manner, as in *Sco'land* for *Scotland*, while *hree* is sounded instead of *three*.

The reasons for differences in intonation have not as yet been discovered. Doubtless climate has much to do with it, but why that should be the case is not known. In the west country, round Glasgow the sentence begins on a high note, and ends on a low one; in the east it is rather the other way. It would be an interesting, if somewhat long and intricate investigation, to tabulate the various varieties of intonation in these islands, and endeavour to evolve some general principle.

According to the preliminary report of the census of 1911, Linlithgowshire has a population of 79,456. Thus, though third from the bottom, or thirty-first in area, it stands twelfth in population, and third in density.

It also shares with Ross and Cromarty, and Stirlingshire, the peculiar distinction of having more males than females; Ross and Cromarty should not be counted, as the preponderance of males was due to the accidental presence of a portion of the Royal Navy. Such a numerical superiority of males is more common among savage tribes where female infanticide is practised or in a new country, than

Dalmeny Village

in long-settled civilised regions. That this is no temporary or even recent feature is proved by the fact that it dates back to the census of 1841, but it must be remembered that in that year the construction of the railway between Edinburgh and Glasgow gave occupation to large numbers of "navvies." There is no doubt that the persistence of the male surplus since then is due to the fact that textile manufactures employing female labour are practically non-existent, and that the main industries are mining and agriculture, in which few, if any, women are nowadays employed. The docks at Bo'ness and the shale retorts also provide work for semi-casual labourers of a low type, who are constantly on the move, and have, therefore, few women with them. The girls whom these various causes render superfluous, become domestic servants, and migrate to other parts, chiefly to Edinburgh.

The population of 79,456 may be divided into inhabitants of municipal and police burghs, numbering 32,520; and of rural districts, numbering 46,936. But there are numerous villages connected with mining, such as Uphall, Broxburn, Dalmeny and Philipstoun, which along with the outlying suburbs of the burghs themselves reduce the number of really rural inhabitants.

As regards density of population, Lanark has 1633 persons per square mile, Midlothian has 1373, and Linlithgow has 657. At the opposite extreme is Sutherland with 10 persons per square mile. The average density for Scotland is 157.

12. Agriculture — Main Cultivations, Stock, Woodlands.

The soil of Linlithgowshire is mainly derived from the boulder clay which covers the greater part of the country with a thick mantle. Near the coast there is an admixture of sand, while along the lower courses of the streams are stretches of fertile alluvium. The coldest soil is in the south and south-west, where lies most of the uncultivated ground, but where the riches underneath compensate for the barrenness of the surface. The most fertile parts are in the midland valley, and in the parish of Dalmeny.

Out of a total land area of 76,861 acres, no less than 57,751, according to the returns for 1911, are under crops and grass. The arable land amounts to 34,393 acres, leaving 23,358 acres for permanent grass. Woods and plantations occupy 4865 acres; mountain and heath land for grazing, 3248. The remaining 10,997 acres consist of waste land and that occupied by buildings.

The farmers of the Lothians have for long been celebrated for their skill and progressiveness. They have not rested content merely with the advantages derived from the natural fertility of the soil, and proximity to the great market of Edinburgh; and they are recognised as the foremost in Scotland, which means the world, for a readiness to introduce new methods, and to utilise the discoveries of experimental science. In 1723 a Society of Improvers was started, in order to consider means of alleviating the long distress and putting an end to the stagnation which dated as far back as the civil wars, and even to the transference of the lands from the Church to temporal owners. Five years later John, Earl of Stair, came to live at Newliston. He introduced "horse-hoeing husbandry," cultivated cabbages, turnips, and carrots by the plough as well as lucerne and sainfoin, and may be regarded as the father of modern farming in the Lothians.

The total area under cereals and pulses is 14,620 acres, 9996 of which are under oats. Wheat covers 2524 acres; barley 1843; rye, beans, and peas are inconsiderable. Small fruit such as strawberries and raspberries account for 53 acres. There are a few small orchards of apples, pears, and plums. Gardening as an occupation is said to have become general early in the seventeenth century, and special mention may be made of the pleasances attached to the various country seats, Hopetoun House, Dalmeny, Dundas Castle, and Newliston.

Turnips, swedes, and potatoes occupy a fair proportion of the area, the former two 3383 acres, the latter 2441 acres. Rape, cabbages, vetches, and mangolds are cultivated in small quantities.

No less than 13,326 acres, or about two-ninths of the whole, are given to clover, sainfoin, and grasses under rotation. As already mentioned, 23,358 acres are under permanent pasture, making in all, three-fifths of the available land devoted to the feeding of animals.

Stock-breeding is not engaged in to any extent, the usual practice being to buy store cattle and sheep, and fatten them for the market. Auction rings exist at Linlithgow and Bathgate; while the renowned cattle market at Falkirk with its annual "tryst" is in Stirlingshire close to the border. 1680 horses are kept for agricultural purposes. The numbers of sheep and cattle, which necessarily fluctuate according to the general character of the weather over the season, were, in 1911, 22,022 and 11,250 respectively. Pigs number only

1985, and even this small amount tends to decrease, owing to the strictness of the public health regulations. Once every cottager kept a pig, but now sties must be removed from inhabited dwellings to such a distance as practically confines them to farms of reasonable size. It is doubtful whether this regulation is an unmixed blessing in the case of many a country cottage, where the sty affords a convenient means of getting rid of kitchen refuse.

Woodlands occupy the proportionately large area of 4865 acres, or one-sixteenth of the whole. The famous trees on the Hopetoun estate have already been referred to, but most of the large landowners prosecute forestry in a scientific manner. Not only are the parks round the country mansions well wooded, but the crests of most of the volcanic hills are picturesquely covered with plantations of coniferous trees.

The number of holdings in the county is 496, with an average size of 116½ acres. More than half range between 50 and 300 acres; nearly one-fourth between 5 and 50 acres; 29 are above 300; and 43 are between 1 and 5 acres. The last are of course beside the towns and are used as market gardens.

In conclusion, it is interesting to note that the average yield in bushels per acre for the ten years 1900 to 1909 was for wheat 43·47; for barley 43·60; and for oats 43·49.

13. Industries and Manufactures.

While Linlithgowshire is in the main a mining and agricultural county, it is not destitute of manufactures or industries, though most of these are on a small scale.

Distilling was at one time much more important than it is now. In Linlithgow alone there were five distilleries. To-day there is only one, producing malt whisky, and it is not constantly working. An interesting feature is the beam-engine built in 1824, which still does all the work of grinding and distilling. The condensing cylinder has never been renewed, yet it is in good order. The works occupy the site of St Magdalene's Convent, which originally belonged to the Lazarites, and was used as a place of entertainment for strangers. Close by is the canal, the water of which is employed for condensing purposes, while the necessary supplies of coal are conveyed along it in barges. Other distilleries in the county are at Bathgate, Kirkliston, and Bo'ness. At Kirkliston the chief output is a malt extract, and only a few hands are employed. The Bo'ness distillery is the principal yeast factory in Scotland. Attention is given to by-products rather than to the spirit itself.

Paper-making, though not on the scale of Midlothian, is of considerable importance. The two mills at Linlithgow produce the fine quality of plate-paper used by the better-class illustrated magazines as well as writing and ledger papers. Alfa or esparto grass is the sole raw material employed. Westfield on the Logie Water, a tributary of the Avon, has also a paper mill.

At Linlithgow is a small glue factory, which has been established over 100 years. The raw material is hide cuttings. The glue is used in the cabinet trade, and much of it goes to London to the makers of pianofortes. A certain amount of grease, which is a by-product, is turned over to the soap manufacturers.

Like many small towns, Linlithgow has a tannery; but the shoemaking trade which was once so important has vanished. In 1793 the Earl of Hopetoun ordered 700 pairs of shoes for his regiment. Many other small industries died out on the introduction of the factory system. Among them were lint and linen yarn mills, tambour factories, and calico printing works.

In 1901 Messrs Nobel built the Regent Factory at Linlithgow. The works cover three acres, and about 200 hands are employed, two-thirds of whom are women and girls. The chief product is safety fuses which are employed to fire shots in mines and quarries. They consist of jute yarn spun round a core of back gunpowder, and waterproofed by a coating of pitch and gutta-percha. This, of course, is only a small part of the Nobel Company's huge business, carried on at Ardeer in Ayrshire and at other places.

Bo'ness has saw-mills, foundries, chemical works, potteries, ship-breaking, ship-building, engineering, and cabinet-making, besides the distillery already mentioned. The saw-mills are largely engaged in making pit-props. At the cabinet-making works doors, windows and shop-counters are turned out successfully in competition with Sweden.

Bo'ness

The Forth Chemical Works produce manures. The raw materials are phosphates from Chile and Tunis, and sulphur and copper ore from Tarsis in Spain. After the sulphur has been extracted the copper ore is sent to Newcastle to be smelted by the Tarsis Company. In 1910 the output of manure was 26,000 tons. About 100 men and 16 women and girls are employed. The various processes are most interesting to study. The copper ore is burned, vapour ascends into a hot tower, where it is mixed with steam. It then passes into a Gay-Lussac cooling tower, where it meets nitrous material, which liberates the sulphuric acid. The latter is then mingled with the phosphates in many different proportions to suit different purposes. The nitrous acids are used over and over again.

Bridgeness Pottery Works near Bo'ness employ 120 hands. The clay is brought in small sailing vessels from Devon and Cornwall; flints come from the north of France; and the stones for grinding the flints from Argyllshire. Dinner, tea, and bedroom ware is produced; and early in 1910 coronation mugs were being made at high pressure. A curious and constant feature is the moulding and decoration of gorgeous china dogs, which are much in evidence in miners' houses.

The shipbuilding industry in Bo'ness is not now important; but it is worthy of mention in connection with Henry Bell, the inventor of the steamboat, who learned his trade with the firm of Shaw and Hart, towards the end of the eighteenth century.

Of the numerous mills marked on the One-Inch

Ordnance Survey Map along the courses of the Almond and the Avon, many have passed out of use. Power derived from water is not to be depended upon in a lowland county. For example, during the long drought of the summer of 1911 West Lothian suffered from a dearth of water for domestic purposes, and the streams were reduced to infinitesimal dimensions. A few grist-mills are still working, as at Kinneil; but the tendency is increasingly towards abandoning the neighbourhood of streams unless coal is easily obtainable.

The last industry we shall mention has quite disappeared, but it has an interesting survival in the Dyers' Corporation of Linlithgow. There are now no dyers in the town, and the activities of the society are mainly social. It has a large benefit fund, and an old-age pension, which all members are bound to receive after they reach a certain age. The Corporation also takes a conspicuous part in the annual ceremony of riding the town's marches. Thus it is one of the old trade guilds which has outlived the trade itself.

14. Mines and Minerals.

Coal, iron, oil-shale, and limestone are the chiet minerals. Coal has been mined from very early times, at least since the end of the thirteenth century. During the reign of James III it was obtained at Bonnytoun, near Linlithgow, by tunnelling into the hillside from the outcrop. At present the industry is carried on round

Bo'ness on the coast, and in the south with Bathgate as a centre. It is more than probable that in the neighbourhood of Linlithgow coal lies beneath the igneous rocks; and as the more easily obtainable fuel becomes exhausted this deposit will be attacked.

At Bo'ness the total thickness of the seams is about 50 feet. The best paying qualities are the cannel and the black-band ironstone coal. The Kinneil Coal Company are also working the easter main coal, which is highly valuable. It is exhausted, however, except under the Firth, and underneath the Millstone Grit, which is above the Carboniferous Limestone series. All the Bo'ness seams deteriorate as they go inland, so that the productive measures do not extend more than a mile and a half from the coast. Out in the Firth they come to an end against the boulder clay which fills up the ancient channel of the river.

The Bathgate and Torphichen Hills seem to have formed an island during the Carboniferous epoch so that the Bathgate coalfield is completely cut off from Bo'ness. At Balbardie the depth of the shaft is 1200 feet, and the principal seam is the cannel coal, which is of great value in making gas. According to Mr Dron, "one ton produces 14,150 cubic feet of gas, of 36-candle power, and 1094 pounds of coke of fair quality."

As one proceeds southwards from the Bathgate Hills the seams increase in number and thickness, but at their best they are inferior to the Bo'ness coal. In the Armadale district the famous Torbanehill cannel coal was much worked. It is now nearly exhausted, but the shafts sunk to work it enable other seams to be exploited, which otherwise owing to their inferior quality would have been left alone. The Torbanehill cannel coal is interesting owing to the number of infusoria found in it, and it is also indestructible by weather agencies.

At Blackburn and near Addiewell seams are worked which also extend into Lanarkshire to the great collieries at Wilsontown. Of these the chief is the main coal.

The total quantity of proven coal in the county is over 200 million tons, of which one-fifth has been worked. Added to that is an estimated quantity of unproven coal, amounting to about the same, and 1000 millions of tons are supposed to lie under the Firth of Forth, making a grand total of fourteen hundred million tons. Of this only about fifty million tons have been used up. It is considered, therefore, allowing for everything, that our supply will last for about 150 years. The cry of exhaustion is no new one. In 1563 an Act was passed prohibiting altogether the export of coal, but it was evaded or ignored. At different times export duties have been put on. Let us again quote from Mr Dron : "Suppose we leave the coal unworked for one year we should lose £25,000,000. If, however, we take the coal, and invest the £25,000,000, at 2½% simple interest it would accumulate in 150 years to £118,000,000." It is a difficult matter, and a controversial one, but it seems that to interfere with a great industry for the problematical benefit of our descendants is, to say the least, a doubtful policy, and very closely resembling the futile act of burying one's talent in a napkin.

East of Bo'ness and Bathgate, the coal seams crop out, and the oil-shale appears at the surface. The group to which these shales belong is known as the Burdiehouse Limestone series, after the famous deposits at Burdiehouse in Midlothian. It is only in Linlithgowshire, however, that the shale is extensively worked, although the formation extends from Berwickshire to Fife. Refineries are situated at Broxburn, Pumpherston, Oakbank, and Uphall; while retorting plants are at work at Dalmeny, Niddry, Hopetoun, Philipstoun, Seafield, and Breich.

The shale is brought up from the mines, broken into lumps, tipped into retorts, and heated. Thus the volatile part is driven off, while the spent shale is conveyed to the "bings" which, to the eye of the casual traveller, form such conspicuous and unpleasing features in the landscape. The yield of the oil-shale is as follows:

1. Permanent Gas.
2. Crude Oil.
3. Ammonia Water.
4. Spent Shale.

The gas is driven back to the furnaces and used for fuel. From the crude oil separate distillations interspersed by washings with oil of vitriol and caustic soda produce a great variety of products, and there is practically no waste after the separation of the spent shale by the first process.

Mr Cadell gives a list which would fill several pages, but we confine ourselves to the most important products. They include, first, naphtha or spirit for special lamps, for dissolving gums, making varnishes, and all sorts of purposes down to removing grease from clothes. Second come burning oils, which owing to their high flash-point have never caused fatal accidents such as frequently occur with American oils. Third are lubricating oils for machinery. Then there is solid paraffin for candle-making. Soft paraffin is almost exclusively used for

Oil-works, Uphall

miners' lamps, owing to its safety flash-point of 353° Fahr. The sulphate of ammonia is sold to farmers for agricultural purposes. It is the best manure for growing sugar-beet. Further, the residues in the form of tars, as well as the crude petroleum, are being increasingly employed as fuel. Their value is one and a half times that of coal; there is no smoke; and storage is easy and safe.

Lastly, there is a growing production of petrol for driving explosive engines and motor cars.

The oil-shale industry dates back to 1850, and from the first it was very successful. In 1877, however, the natural petroleum of Pennsylvania came into competition, with the result that the price eventually dropped to about one-third of its former amount. Nothing daunted, the refiners tried all possible means to increase the efficiency of the works, and directed their attention especially to the by-products. The result is that the amount per ton of saleable material has been with some products trebled. It is a remarkable testimony to the fact that British industry is far from dead. And the extension of the use of oil-fuel by war vessels will in time enable the oil companies to reap the full reward of their enterprise. The production in 1909 amounted to 2,261,086 tons, worth £621,799.

The existence of black-band ironstone west of Bo'ness led to the establishment of iron furnaces. The most profitable seam was one of about twelve inches thick resting on the top of the cannel. This has been worked out, and the furnaces abandoned. But there is still at Bo'ness a foundry which manufactures hoes and spades. In the Armadale district an erratic ironstone is found, varying in width from two inches to seven feet. But there are nowhere such extensive deposits as those which originated the famous Carron Ironworks, a few miles farther west in Stirlingshire.

No precious metals are now worked in the county. In the reign of James VI much excitement was caused by the news that a collier, named Alexander Maund, had picked up a stone containing silver. Work was at once begun under lease. It was said that 24 ounces of silver were gained from each hundredweight of ore. One man was reported to have got £100 of silver in a day. Captivated by this, James took over the mine himself. Extensive plant was obtained, and workmen imported from abroad; but it proved a failure. The mine was again leased with the same result. The only person who did not lose by it was Sir Thomas Hamilton of Monkland, the King's Advocate, and the first lessee, who got £5000 for surrendering his claims. The site of the mine was a little to the east of Cairnpapple Hill near Linlithgow. The adventure is remembered now by the names Silvermine transferred to a neighbouring quarry, and Silvermills in Edinburgh, where the ore was treated.

Limestone has long been worked in the county. It was formerly used in agriculture, and the limekiln was a familiar object. Now it is chiefly employed as a flux in iron smelting. The Burdiehouse limestone, which contains over 94 per cent. of carbonate of lime, is found chiefly in Midlothian; but there is no doubt that extensive deposits lie under the shales. The Great Limestone formerly worked in the Silvermine quarries varied from 25 feet to 39 feet in thickness.

Near Blackburn "lakestone" or picrite occurs in connection with the coal measures. It weathers very quickly, but it has great capacity for resisting heat, and it is used for making soles for bake ovens.

The dolerite sheets at Bo'ness and West Craig near

Blackburn, and the columnar basalt near Linlithgow, are quarried for road-metal.

Freestone of inferior quality is worked at Bo'ness, and a soft stone at Cauldhame. The Binny quarries are at present closed. The sandstones associated with the oil-shale have long provided a beautiful and durable stone for building purposes. The quality is similar to that of the now exhausted quarries at Craigleith near Granton, which along with Binny have supplied most of the material for the new town of Edinburgh.

The glacial and alluvial deposits of the Great Ice Age have here as elsewhere provided material of economic importance. At Blackness the old 100 foot sea-beach has cut into the till, from which clay for making bricks and tiles was obtained. At Winchburgh and Ecclesmachan and at Foulshiels Colliery the boulder clay and "blaes" or argillaceous shale are ground in a pug-mill and passed through brickmaking machines. Bricks and tiles are also made from alluvial clay west of Bathgate.

A deposit of shell-marl in Linlithgow Loch is occasionally worked by farmers.

15. Shipping and Trade.

Although Linlithgowshire presents a front to the estuary and Firth of Forth, the fishing industry is practically non-existent. The main railway lines are all inland except at Queensferry. Even there, the station is one mile distant, and the road to it goes up a steep hill. Cramond is two miles, Blackness four miles, from the nearest railway station, while Bo'ness is the terminus of a branch single line. Any locally owned boats, therefore, use more eastern ports, such as Newhaven, which has the large market of the metropolis close at hand. A few boats may be observed at Cramond; but the difficulty of access by the winding channel of the Almond over Drum Sands forbids further development.

As its name implies, Queensferry is more of a ferry-port than a seaport, and the importance of the ferry has greatly diminished since the opening of the Forth Bridge. A pier projecting from the middle of the town is used by a fleet of pleasure steamers. Farther west is the enclosed basin of Port Edgar, said to be named after a brother of Margaret, Malcolm Canmore's Queen. To it descends a railway, a survival of the original line from Edinburgh. Though no longer open for passenger traffic, it is used in connection with the oil-shale mines in the vicinity. At Port Edgar, George IV embarked on his return to England in 1822.

The glory of Blackness has long departed. One of the most ancient ports in the kingdom, and, while Berwick-on-Tweed was in the hands of the English, the first for amount of trade, it has succumbed from natural causes. It was originally the port of Linlithgow, and was completely under subjection, even the Custom House being in the county town. All ships had to be beached at high water, and loaded or unloaded at low water; traces still remain of the paved causeways used by horses which conveyed cargoes to or from the vessels. From the time that ships began to be built of a larger size than herring

Bo'ness, from the Harbour

boats, Blackness was doomed. Finally the Custom House was removed to its western rival Bo'ness, and the port practically closed. A short iron pier projects from the point, but that is for the exclusive use of the powder magazine and factory in the Castle.

Bo'ness, therefore, is the only port in the county worthy of the name. From 1750 to 1780 it was the third seaport in Scotland. To-day it occupies a much more humble position, not because its trade has declined, but because it has been outstripped by successful rivals. A heavy blow was dealt by the opening of the Forth and Clyde Canal early in the nineteenth century, and by the erection in 1810 of Grangemouth as a separate port. Within recent years dock accommodation at Grangemouth has been greatly extended, and now it somewhat overshadows the older port. Further, the situation of Bo'ness unfits it for serving more than merely local needs. Placed on the narrow platform of an ancient sea-beach, and cut off from the south by a steep ridge, it lacks the "hinterland" essential for the development of a large port. Then the railway leading inland to Manuel on the main line and on to Bathgate is a single track, and this is fatal to really large traffic. Thirdly, there is no coast road running eastwards. Thus one cannot expect for Bo'ness any great growth in the future.

In spite of natural disadvantages, the trade of the port is considerable. This is largely due to the neighbouring collieries. In 1881 great improvements were completed. The quays were rebuilt, a new pier erected, and the dock accommodation extended by 7½ acres. The length of quayage in dock and harbour is now 4300 feet. The depth of water on the sill of the harbour is 22 feet at spring tides, and 18 feet at neap tides, but an allowance of two feet must be made for silt in the channel. The newest hydraulic machinery is now available, chiefly for handling coal. There are three coal-shipping hoists, each capable of loading 2000 tons in 24 hours, and several portable cranes. Large sheds are provided for storage of cargoes.

The imports are timber, pit-wood for the mines, esparto or alfa grass for the paper mills, grain, iron-ore, copper-ore, and phosphate for the chemical works, clay, grindstones and flints for the potteries. Exports are composed of coal, iron, copper-ore freed from sulphur, and general goods. Steamers sail once a week direct between Bo'ness and London. The value of the imports in 1910 was £310,446; the exports in 1909 amounted to £342,049. Customs revenue in 1910 came to £3603. The quantity of coal dealt with was 880,297 tons.

The following tabular statement may be useful. The figures are for 1910.

		No. of vessels	Tonnage
General Coasting Trade			
Entered		493	142,999
Cleared		545	160,487
	Totals	1038	303,486
Foreign Trade			
Entered		561	344,265
Cleared		589	329,784
	Totals	1150	674,049
	Grand Totals	2188	977,535

16. History of the County.

Linlithgowshire was always debatable ground, and from the place-names we may infer that West Lothian was the battlefield of Picts, Scots and Britons. One of King Arthur's battles against the Angles took place in 580 at the Mons Badonis, which by some authorities has been identified with Bowden Hill, near Linlithgow. In 638 Domnall Breac was defeated by the Angles at Glenmuireson, probably, says Skene, the Murieston Water, which flows through Midlothian into the Linhouse Burn near Midcalder.

But the earliest authentic reference to the county mentions the establishment of a monastery at Eoriercorn, Aebbercurnig, or Abercorn, under Bishop Wilfrid about the year 680. This is interesting as showing an advance northwards of the Northumbrian Church, which followed Rome, into the territory of the Celtic Church of Iona. It was a sign of the steady progress of the Angles to the Forth. Their political hold on West Lothian was never very strong, though they gradually displaced the Celtic population. The historical events of the county cluster, for the most part, round the county town. David I probably resided at Linlithgow for the chase, and built the manor-house of wood or stone which was the forerunner of the palace. His grandsons, Malcolm IV and William the Lyon, certainly lived there at times.

Edward I of England had a close connection with the burgh. During the winter of 1301–1302 he built a stockade and dug a ditch round the manor-house, and enclosed a considerable area, which to this day is crown property, and called the Peel of Linlithgow. A garrison was left there, but it was surrendered soon after Bannockburn, and the stockade and house were destroyed by the Scots.

Robert II, the first of the Stuarts, was elected and proclaimed at Linlithgow. In 1388 he elevated Linlithgow into a royal burgh, leasing its customs and those of the port of Blackness to the burgesses for the sum of £5 yearly. It is only now, towards the end of the fourteenth century, that we begin to have clear glimpses of the condition of the town and county.

Wood was universally used for buildings and thatch for roofs except possibly for the manor-house. Hence great damage was frequently caused by fire, but this damage was also easily repaired. The English burnt the town in 1337, 1411, and 1424.

In 1453 Blackness Castle was besieged. The late owner, Sir George Crichton, had left it by will to the king. His son and heir naturally objected, and held out for a fortnight, but surrendered on being promised other lands in compensation. Since then Blackness Castle has been crown property. In the wars with the Douglases James captured and destroyed their towers at Inveravon and Abercorn. In these sieges a bombard was used, probably Mons Meg, still preserved in Edinburgh Castle.

Henry VI of England, after his defeat at Towton in 1461, took refuge in Scotland and resided in Linlithgow Palace for some time. In 1466 while James III was at the Palace he was seized and carried off by Lord Boyd, an act which though successful for the time, caused his speedy fall when James was able to assert his power. The artist king was fond of Linlithgow and spent a considerable sum in repairing and decorating the palace. It was at Blackness that the trouble began with his nobles. There James met the rebel forces who had possession of his son, and a hollow peace was agreed to. A month later he was murdered after the battle of Sauchie and the victorious barons celebrated their victory at Linlithgow.

The two succeeding kings were the last to spend any considerable time at Linlithgow, and the town prospered greatly during their reigns. James IV regularly passed Christmas and Easter there, amid great festivities. The most notable event was the appearance of the "ghost" in St Michael's Church, warning the king against going to war with England. The supernatural visitor was probably an emissary employed by Queen Margaret, but, as we all know, the device was unsuccessful, and James went south to meet disaster and death at Flodden[1].

For some years the palace was neglected, but in 1525–1526 James V spent a night there, and with Douglas celebrated with unseemly levity his victory over Arran. A few months later Arran and Angus, now in alliance, defeated an army led by the Earl of Lennox, who acted for the Queen Dowager along with Chancellor Beaton. Lennox attempted to cross the Avon near Manuel, but

[1] See the graphic account of the apparition in Pitscottie's *Chronicles*, and in Scott's *Marmion*.

his army was completely routed and he himself slain. This is usually called the battle of Linlithgow Bridge.

James V

When James became king in fact as well as in name, he frequently resided at Linlithgow. The money required for alterations in the palace was obtained partly by fines and confiscations from Protestants. The first Scottish martyr for the Reformation was Patrick Hamilton of Kingscavil; and Henry Forest, a native of Linlithgow, soon followed him. It was in the Peel that Sir David Lyndsay's *Satire of the Three Estates* was performed in January, 1539–40. Mary Stuart, Queen of Scottish hearts, was born in Linlithgow Palace in 1542. During her absence in France, the first Regent, the Earl of Arran, lived at the palace. It was more remote from England than Edinburgh. Another reason was that Linlithgow was singularly free from the plague which at that time was raging in Scotland. The "hollow" was always a healthy place, and there are frequent references in history to the migration to Linlithgow of the government of the day in order to avoid the plague. This occurred several times during the next hundred years. Parliament and the Lords of Session met in the great hall of the palace; in 1646 even the professors from Edinburgh University taught their classes in the parish church.

In 1559, when the Queen Dowager was Regent, the Lords of the Congregation "cleansed" St Michael's Church, and probably the palace, of images and altars. They left the image of St Michael in the church because it could not be removed without injuring the building. The statues of the Virgin Mary and of Pope Julius II were left in the palace. The latter is said to have been destroyed afterwards by a blacksmith in an excess of zealous frenzy.

The only important later connection of Queen Mary

with the county was her "capture" by Bothwell. Some authorities place the scene at Linlithgow Bridge over the

Earl of Murray Tablet, Linlithgow

Avon, others at Cramond Bridge over the Almond. Wherever the spot was, this event brought Mary's affairs to a climax and was the immediate cause of her ruin.

Her last visit was after her escape from Lochleven, when she spent a night at Niddry Castle.

The regency of the Earl of Murray came to an end on January 23, 1569–70. As he was riding through Linlithgow, Hamilton of Bothwellhaugh shot him from the window of a house on the south side of the High Street. The Sheriff Court House now stands on the site of the house. After Murray's death Linlithgow became the headquarters of the Queen's party, and was exposed to the attacks of the English. An army under Sir William Drury burned the mansions at Kinneil, Kingscavil, Binny, and Park. Linlithgow itself was doomed, but was saved by the Earl of Morton and by the petitions of weeping women.

Three visits of James VI are worthy of mention. In 1585 Parliament met in the Great Hall and agreed to a league with England. Eleven years later the Protestant ministers preached against Episcopacy and a tumult arose in Edinburgh. James came to Linlithgow, threatened to remove to it the Law Courts and Parliament, to destroy Edinburgh and sow its site with salt. The capital was not proof against this awful threat, and made complete submission. The third visit was in 1617, when James was graciously pleased to revisit his native land. Unheard-of preparations were made to receive him. An address composed by William Drummond of Hawthornden was spoken by Wiseman, headmaster of the burgh school, literally out of the mouth of a lion.

> "The king of beasts speaks to thee from his den,
> Who, though he now enclosëd be in plaster,
> When he was free, was Lithgow's wise schoolmaster."

During the reign Blackness Castle and afterwards Linlithgow Palace were used as prisons for recalcitrant Presbyterian ministers. Six of them, including John Welsh and John Forbes, were condemned by a packed jury in the Tolbooth, Linlithgow. They were ably defended by Thomas Hope, the founder of the fortunes of the present family of Hopetoun.

Great activity preceded the visit of Charles I in 1633. Fords and bridges were repaired, roads were widened. No less than 600 horses and carts were ordered to be provided from the various parishes for the carriage of the king's luggage. After all, Charles stayed only one night, the last that a king ever spent within its walls. The burgh received confirmation of its ancient charter, but the expenses were ruinous.

During the Civil War Charles tried his father's plan of threatening to make Linlithgow the capital instead of Edinburgh, but this time without success. After Philiphaugh Parliament did meet there, and it is said that 80 women and children were thrown into the Avon at Linlithgow bridge and drowned, an echo of the wholesale slaughter of the prisoners on Slain Men's Lea below Newark Castle.

Charles II was proclaimed king at Linlithgow Cross on the 7th of February, 1649, that is, immediately on the news of his father's execution reaching Scotland. Eighteen months later Cromwell came after the victory at Dunbar, destroyed many houses in the Kirkgate, including the Pre-Reformation manse, and used the stones to build a wall round the palace. The Town Council fled to Culross, where they remained during his stay. On his departure General Monk came and resided in the Palace until he left on his march to England in 1659.

Great rejoicings took place on the Restoration. A huge bonfire was built, into which were thrown all available relics of the hated Commonwealth, including Covenants and Presbyterian books. Cromwell's wall was taken down. James, Duke of York, afterwards James VII, lived in the palace while he was controlling affairs in Scotland towards the end of the reign of Charles II.

During the '15 Linlithgow, true to its traditions, was strongly Jacobite. Its loyalty being suspected owing to the known opinions of James Glen the provost, a Dutch garrison of 6000 men was maintained in the palace and burgh. To meet the large expense thus caused to the burgh, an extra tax of twopence per pint was laid upon ale and beer consumed in the town. The keeper of the palace, James fifth Earl of Linlithgow, a strong Jacobite, had his estates confiscated; and the earldom became extinct.

In 1745 Linlithgow was occupied by Colonel Gardiner with his dragoons. On the approach of the Chevalier, he withdrew, and Charles entered the town without opposition. He was entertained in the palace by Mrs Glen Gordon, who revived the ancient glories by making the beautiful fountain in the courtyard run with wine. Charles spent the same night at Kingscavil House, while his army bivouacked at Threemiletown.

Next year General Hawley ordered the town to prepare provisions for his army. Lord George Murray, who lay at St Ninian's, marched over and seized them, exchanging courtesies with the van of the Hanoverians at Linlithgow Bridge. This was the last approach to a conflict in the county. After the battle of Falkirk Hawley retreated in disorder to Linlithgow. Later the palace was occupied by the Duke of Cumberland.

In 1859 the burgh had a law-suit with the North British Railway Company. The Town Council based on their ancient charter a claim of charging dues on goods carried by rail. The case was lost and the town became bankrupt. One curious result was that the corporation seats in the parish church were sold by public auction.

The story of Linlithgowshire to-day is one of peaceful industrial development. It does not seek to live on memories of the past. Linlithgow rightly will never forget that she was at times the capital of Scotland. Should the palace ever be restored as has been proposed, she might again become a favourite residence of kings. This hope is doubly pleasing to the inhabitants of the county town, which is somewhat overshadowed by its thriving neighbours, Bo'ness and Bathgate.

17. Antiquities—The Roman Wall.

Prehistoric relics in general are divided into two classes, the unpolished or Palaeolithic, and the polished and more highly finished or Neolithic. Authorities are almost unanimous in the opinion that Palaeolithic man did not reach Scotland. In southern England the greater number of the prehistoric implements are of flint, which is plentiful among the chalk, and which lends itself to splitting or flaking for the purpose of producing sharp edges or points. In Scotland, however, flint is very rare. None exists in the Forth area, hence the weapons and tools have been constructed of the material ready to hand, such as trap rock. No other relics which can certainly be ascribed to Palaeolithic or Neolithic man have been found.

When the Celts first settled in the county, the low ground was probably marshy and covered with forest. Hence any traces of early occupation appear now on the ridges, and particularly on the intrusive bosses between Bathgate and Linlithgow. Subsequent cultivation must have long removed any remains in the valleys or along the shore, if such ever existed. Those primitive people built round walls of stone, possibly strengthened with timber, and turf or sods, within which were their rude huts of branches. In these enclosures they even kept their cattle and sheep. Most of the structures marked "Fort" in the Ordnance Survey Map are of this nature, and it requires a considerable effort of the imagination to recognise their crumbling remains. On Cockleroy, Torphichen, and Cairnpapple Hills are traces of "forts" or earth-works.

It is characteristic of mankind to set up monuments, which in their rudest form consist of mounds of earth or of loose stones. An advance is indicated by the erection of columns of rock or "standing-stones"; sometimes two stones close together are crowned by a third laid horizontally along the tops, forming a rude arch. The latter is called a cromlech, an example of which stands on

Cockleroy near Kipps. About a mile east of Bathgate on the road to Bangour is a group of standing-stones. There is no record of human remains having been found in connection with any of those relics of antiquity. Near Kirkliston, on Westbriggs farm, stands a stone called the "Catstane." It is 4½ feet above the ground, and 11½ feet in circumference. On it is an inscription, which Mr Gillespie in *Round about Falkirk* gives as INOCT · VMVLO · JACI · VETTAD · VICTA. These are the only letters decipherable, and their reference is unknown.

But the chief memorial of antiquity in the county is the remains of the Roman Wall. Agricola, after his campaign in the north, constructed a chain of forts from the Forth to the Clyde connected by a rampart of earth; but this was abandoned in 85 A.D. Sixty years later, in the reign of Antoninus Pius, the tribune Lollius Urbicus carried out the famous work, which can still be traced at many points along its length. Except for an interval of three years, from 155 to 158, this wall was effectively occupied till about 183, a period of forty years in all.

Although the word "wall" is used, it must not be imagined that it was composed of dressed stones like the more celebrated and better preserved wall stretching from the Tyne to the Solway. The foundation consisted of roughly dressed kerb-stones, 14 feet apart, the interval being filled with rough stones placed close together. The material was freestone, which had in some places to be conveyed from a considerable distance. The superstructure consisted of layers of turf, the whole rising to a height of 10 feet. On the top was possibly a wooden palisade. Every two miles or so were forts of the usual square shape, built against the south side of the rampart. Through them ran the military way, which was not a regularly paved road, but had a foundation of large stones, covered with smaller stones rising to a crown in the centre. On the northern side of the rampart was a V-shaped ditch, which was in some parts rounded at the bottom instead of pointed. It was about 20 feet deep and 40 feet broad. The rampart is a little way down the northern side of a low range of hills, while the road keeps close to the crown of the ridge.

It is a common misconception that Roman roads and walls ran in a straight line regardless of physical obstacles. Anyone who has visited and studied the Forth and Clyde wall or the Tyne and Solway, knows that such is not the case. In a frontier region the Romans placed their defensive works a little down the outer slope while the road followed the crest of the ridge. The object seems to have been that troops on the march had a double view of the country and were thus secure from surprise, while, in the event of an attack on the wall, the enemy had to charge uphill and the defenders had the advantage of a commanding position. Thus wall and road take a sinuous line corresponding to the trend of the ridge.

The Forth and Clyde wall and road extended from Bridgeness to Old Kilpatrick, a distance of 36 miles. Only here and there can traces be seen of the road. It sometimes appears as a grass track across fields, at other places it is used by a modern highway. Its foundations, as well as those of the rampart, have been frequently turned up by the plough, or revealed by drainage operations. Doubtless much of the stone has long been removed and used in boundary walls of fields, and even in buildings.

Although the line of the wall is confidently marked on the Ordnance Survey Map, the only portions in Linlithgowshire which are indisputable are at Inveravon, White Bridge, and Graham's Dyke Road, with a total length of five-eighths of a mile. Kinneil means Wallsend, but it is known that the rampart extended three miles farther east to Bridgeness. The name Walton occurs about a mile and a half inland from Bridgeness, which has led to the speculation that the wall continued to Cramond, but there is no proof whatever to support the assertion. At Walton there may have been a camp of observation, but nothing except the name remains. No vestiges are left of the forts which most probably existed at Inveravon, Kinneil, and Carriden.

Graham's Dyke Road, just outside Bo'ness, is an interesting case of the survival of an ancient place-name. Who Graham was we cannot tell, but the word "dyke" irresistibly suggests the wall. In 1649, when Bo'ness was separated from Kinneil and erected into a parish, Graham's Dyke is mentioned in the Act of Parliament as the southern boundary.

So far no relics have been found in the county comparable in number to those unearthed in the forts to the westward. A gold coin of Vespasian was dug up at Carriden, but no pottery or implements have been discovered like the treasures from the wells or rubbish-heaps elsewhere. One object, however, of great importance, found at Bridgeness, is now in the National Museum of Antiquities in Edinburgh[1]. It is a "distance slab" set up to commemorate the completion of a portion of the *limes*. It appears that the work was divided into sections, and distributed among the various legions. The wall was probably begun at the Clyde end and carried across the country to the Forth. If this was the case, the completion of the task would undoubtedly be made the occasion for a festival of a semi-religious character, and the slab bears witness to this fact. It is nine feet two inches long, and three feet eleven inches high. It is decorated with carvings, which were probably coloured, and bears a well-preserved inscription. The principal features are three in number. A horseman, spear in hand, is represented charging four naked Caledonians. Another panel displays the *suovetaurilia*, or preparations for sacrifice. The animals to be sacrificed are advancing left towards a figure seated right; behind is a youth, the tibicen, playing a double flute; in the background are five male figures, the foremost holding a *patera* over the altar, one is behind a standard with the letters LEG
II
AVG.

The principal inscription is:

IMP · CAES · TITO · AELIO
HADRI · ANTONINO
AVG · PIO · P · P · LEG · II
AVG · PER · M · P IIII ƉCLII
FEC

[1] See *The Roman Wall in Scotland*, by George Macdonald, M.A., LL.D.

The "Distance Slab" from the Roman Wall

The slab is a record of the completion by the Second Legion of 4652 paces of the *limes*. It is also an evidence that the occasion was seized to celebrate the conquest of the Caledonians and the conclusion of the laborious work on the wall by a *lustratio* or ceremony of purification. Curiously enough the slab was found face downwards, showing that on the abandonment of the wall it was placed thus in order to preserve it against the day of a possible return of the Roman legions.

18. Architecture—(*a*) Ecclesiastical.

In the early days of the Celtic Church, religious communities gathered together in a collection of rude huts of wattles, bee-hived in structure, while the solitary ascetic made holy by his prayers the dank cave in the wind-swept cliff. The dampness of the climate, especially in the western isles, soon led to the substitution of stone for boughs, though the bee-hive shape was retained. The place of lime was supplied by turf or moss, and the buildings were consequently very diminutive in size. Saxon and Norman influence began to be felt in the eleventh century, and during the next two hundred years there was a gradual development of the Romanesque or Norman style. This was marked by the rounded arch and tower, and the introduction of ornament. From the thirteenth to the sixteenth century the Gothic or Pointed architecture was supreme. This style had three stages—the First Pointed in the thirteenth century, the Middle

Pointed or Decorated in the fourteenth and fifteenth centuries, and the third or Late Pointed period. While the Perpendicular or Late Pointed prevailed in England and the Flamboyant in France, in Scotland there arose a style peculiar to itself. During the sixteenth and seventeenth centuries the Renaissance influence slowly

Dalmeny Church

penetrated, but it competed with the Tudor style from England and the surviving Gothic. At the present day churches are built in many different and even mixed styles, but modern ecclesiastical architecture has as yet not developed any distinctive style of its own.

The parish church of Dalmeny is the finest example of Norman ecclesiastical architecture left in Scotland.

The south front and east end are of twelfth century work, and the doorway with its lovely arch is unequalled in the country. A gallery, porch, and staircase have been added on the north side, but the original interior is plainly distinguishable. It consists of a nave and a chancel with an eastern apse. Between them is an elaborate chancel arch. The church has recently been restored, unsightly modern pews removed, and by turning one's back on the north wing, one may see it practically in its original form. It is a monument of the Norman style in Scotland, which, happily, will stand for a long time to come.

Most of the other parish churches in the county are interesting, but none are so completely Norman as Dal meny. Kinneil Church is now a ruin covered with ivy, and but little of its structure can be seen. Another ivy-clad ruin is the former parish church of Bathgate, a doorway being just distinguishable. Uphall Church was Norman, but nothing remains of it except a bell dated 1441 which is in the present modern building, and a font which has found its way to the Roman Catholic chapel at Broxburn. The name Uphall dates from the Reformation, the parish having been originally named Strathbrock, "the vale of the badger."

Abercorn Church has been so much altered that only a doorway remains of the original structure. In the churchyard are two solid hog-backed tombstones, their sides carved with scales, which must be of early date.

The old name of Kirkliston was Temple Liston, showing that there the Knights Templars had an establishment, no traces of which are now visible. It was subsidiary to their chief possessions at Temple in Midlothian. The church of Kirkliston is a fine building with several features dating back to Norman times. The tower, 21 feet square, with solid buttresses, indicates the Transition or First Pointed period. A square projection contains a staircase on the wheel pattern. The windows are extremely narrow. In the south wall is a beautiful

Torphichen Preceptory

Norman doorway which has been partly restored. The belfry is seventeenth century work.

Another semi-religious order—the Knights of St John of Jerusalem, or the Hospitallers—has left a monument at Torphichen. It resembles a castle rather than a monastery, and there is even a dwelling-house over the transepts. The nave has been restored, and is used as the parish church. In 1168 it acknowledged itself a daughter of St Michael's at Linlithgow.

On the dissolution of the Templars in 1312, their property was transferred to the Hospitallers, who in their turn were dispossessed at the Reformation.

The Carmelite Friars had two establishments in the county, one at Linlithgow and one at South Queensferry. The former has utterly disappeared, having been pulled down, despite some efforts to save it, to make way for modern buildings. Of the South Queensferry monastery practically nothing remains; but the church was restored by the Dean and Chapter of St Mary's Episcopal Cathedral in Edinburgh, and is used as a place of worship. Its earliest charter is dated 1457, but it was founded and endowed by Dundas of Dundas in 1338.

St Michael's Parish Church of Linlithgow was a mother-church in the reign of David I. In 1242 a new building was consecrated by the Bishop of St Andrews, which suffered damage on several occasions when the town was burned by the English. The present church dates from the beginning of the sixteenth century. It has been repaired from time to time; and a careful and reverent restoration was completed in 1896. It is one of the finest examples of Gothic architecture in Scotland. The length from east to west is 187 feet, the breadth, including the aisles, 105 feet. The square tower at the western end had formerly a crown like that of St Giles in Edinburgh. About 100 years ago it was considered unstable and was ordered to be taken down. The mason who had the contract discovered that it was perfectly

St Michael's Church, Linlithgow

sound; but in order to make sure of his money, he completed the destruction before publishing his information.

Most of the modern churches in the county are handsome buildings, but few are of special interest. The most striking is Bathgate parish church, completed in 1884. It has a square tower 108 feet high. The east gable is relieved by a semi-circular apse, pierced by five memorial windows.

19. Architecture—(*b*) Military.

The early Celtic architecture in stone was a round wall or "broch" enclosing a court open to the sky. The wall was of "dry stones," for lime did not come in till Norman times. In the thickness of the wall were the staircases and rooms—a feature which was imitated in the castles of later days. After the War of Independence there was a complete change. Strong square towers with flat and battlemented tops, and arrow-slits for windows, were built in commanding situations. Gradually a courtyard was added, enclosed by a wall. The wall gave place to buildings, stables and living-rooms. When more room was required, the early plan was to add another storey or two; later on, a wing was thrown out. Hence the L, T, E, and Z plans of more secure days. Another development was the substitution of a balustrade for a wall, which is the best mark of the abandonment of military for domestic architecture.

No elaborate defensive works seem ever to have existed

round any of the towns in West Lothian. Linlithgow had a wall of rough, undressed and unmortared stones, with gates or "ports" at the east and west ends, but it was not strong enough ever to resist an organised attack. The villages indeed were hardly worth defending as the houses were built of wood and turf, thatched with heather or rushes. On the approach of invaders the inhabitants

Blackness Castle

fled, returning to their smouldering hovels after their enemies had retired. Such poor shelters were easily renewed.

The sole purely military castle in the county is Blackness. It is situated at the sea-end of a narrow promontory, and therefore required strong defences only on one side. The oldest part is an oblong "tower or keep with a circular staircase tower at the north-east angle." It is surrounded by a wall with a thick parapet pierced by large port-holes. As the castle is crown property, and being used as a powder magazine and factory, is occupied by a garrison, it is closed to visitors.

Dundas Castle

It is possible to walk round the castle at low water.

Blackness was originally the property of Sir George Crichton, who left it by will to James II. It was damaged by fire in 1465, and threatened by an English fleet in 1481. In Covenanting times, it was used as a prison for recalcitrant ministers. Its remote situation makes it very suitable for its present use as a powder factory.

Near Blackness is Dundas Castle, which was erected in the first half of the fifteenth century. The masonry is still in a good state of preservation. About the beginning of the eighteenth century the lower storeys were fitted up as a distillery, and much of the brick-work still remains in the interior. On the tower was a beacon which could be seen from Blackness. The castle was on the L plan, and a wing was afterwards added on the north-west. In the pleasure ground are a fountain and dial of Renaissance work, the former being a fine work of art.

Another building, part of which belongs to the same period, is Kinneil Castle above Bo'ness. The oldest portion is an oblong keep, which was probably built in the fifteenth century. It is an ancient seat of the Hamilton family, and was enlarged from time to time, but was converted into its present form by Duke William in the reign of Charles II. From its association with James Watt and Dr Roebuck, it is one of the most famous houses in the county. Unlike Dundas Castle, it is still habitable.

Three miles east of Linlithgow is Niddry Castle, home of the "loyal Seatons," built by George, 4th Lord Seaton, in the reign of James IV. It is now ruinous and roofless, but noted architecturally for the unusual design of the pointed pediments above the windows.

Ochiltree Castle, three miles south-east of Linlithgow, is mentioned for its remarkable doorway on the west side.

It was built about 1560, and was enlarged by lengthening one leg of the L in 1610.

Bridge Castle in Torphichen parish is also on the L plan. The oldest part belongs to the first half of the sixteenth century. About fifty years ago it was completely reconstructed, and it is now a commodious mansion house.

20. Architecture — (*c*) Municipal and Domestic.

As there are only two royal and three other burghs in Linlithgowshire, we cannot expect to find many public buildings. Armadale, Bathgate and Bo'ness have municipal buildings, Broxburn has a public hall, but in harmony with the industrial activities of these towns, utility rather than ornament has been the guiding principle.

Linlithgow Town House was built by Sir Robert Miln of Barnton in 1668. The curious portico in front gives it a somewhat Italian appearance, and it has been ascribed—needless to say, with but slight foundation—to Sir Christopher Wren. The Cross Well, re-erected in 1807, is a copy of one which dates from the reign of James V. It is a notable example of grotesque Gothic carving. The Sheriff Court Rooms, County Hall, and Offices are housed in a plain building of various dates, erected on the site of the house whence Bothwellhaugh shot the Regent Murray. In the wall is a tablet commemorating the incident, with the curious mistake of giving the 20th instead of the 23rd of January as the date of the assassination. The Victoria Hall on the north side of the High Street and opposite the site of the ancient cross, is, as its name implies, a modern building. The style is Scottish Baronial.

South Queensferry possesses a quaint Tolbooth with a slated spire, but it has been disfigured by the insertion of a "Jubilee" clock.

The finest private mansion in the county is Hopetoun House, the chief seat of the Marquis of Linlithgow. Planned by Sir William Bruce of Kinross, it was begun in 1696, and completed by William Adam. Situated near the edge of a bluff above the estuary, it commands a glorious view down the Firth, and from it the Forth Bridge, distant two miles, acquires a grace which at close quarters would seem impossible. The classical style of the mansion is in perfect harmony with the noble trees, and the almost southern softness of the landscape. Close by is Hopetoun Tower, of octagonal shape and a relic of an L plan house.

In a lovely dell near Abercorn is Midhope, which was inhabited in the seventeenth century by the Earls of Linlithgow while they were keepers of Linlithgow Palace. It is still perfectly preserved, and used as a dwelling-house for pensioners of the Hopetoun family. The oldest part is a block at the west end, six storeys in height; the east wing has four storeys; and a newer wing in Renaissance style is smaller. The plan of the whole is oblong with a courtyard on the south front.

Another famous seat is Dalmeny House, one of the residences of the Earl of Rosebery. The mansion has

Hopetoun House

a pleasant situation on a terrace a few hundred yards from the Firth, but there is nothing remarkable about the architecture. Close by on a projecting point is Barnbougle Castle, which is almost entirely modern. The original building dates from the seventeenth century, but only a portion of the north wall is left, which forms

Midhope Castle

part of the present structure. The grounds come close to the sea, and are protected by a wall surmounted by a balustrade, which may be regarded as a conventional survival of defensive fortification. This is an early example of the transition from semi-military to purely domestic architecture.

Many of the other ancient mansions in the county are now in ruins. They have passed through the stages of occupation by the family, gradual neglect, transformation into farm-houses, or lodgings for farm labourers, and finally decay and abandonment. Amongst these are Kirkhill, near Broxburn, of which all that remains is a window with the Erskine coat of arms over it built into

Dalmeny House

a farmhouse; Duntarvie Castle, near Winchburgh, where the Marchioness of Abercorn was imprisoned for nonconformity in 1628; Bonhard, near Linlithgow, which belonged to the ancient family of the Cornwalls; and Kipps House, the family seat of the Boyds.

Some, however, are still in occupation. Grangepans, near Bo'ness, built in 1564, was long owned by a branch of the Hamilton family. In 1790 it was purchased by John and William Cadell, who with Dr Roebuck established the Carron Iron Industry near Falkirk. The house is on the T plan with curious "angularly-placed and detached chimneys." Houston House near Uphall is of the seventeenth century type, and has been transformed from a keep into a mansion.

There is no lack of building material in the county, both sandstone and limestone being abundant. Some granite has been imported from Aberdeen in modern times for ornamental purposes. But the local sandstone (except at Bo'ness, where it is rather soft) is both firm and durable, while it is easily cut.

21. Architecture—(*d*) Linlithgow Palace.

"Of all the palaces so fair,
  Built for the royal dwelling,
In Scotland far beyond compare,
  Linlithgow is excelling;
And in its park, in jovial June,
How sweet the merry linnets tune
  How blithe the blackbird's lay!
The wild-buck bells from ferny brake,
The coot dives merry on the lake;
The saddest heart might pleasure take
  To see all nature gay."

SCOTT, *Marmion.*

From the days of David I Linlithgow was a favourite place of residence for Scottish sovereigns. Beginning with a hunting lodge probably built of timber, the kings

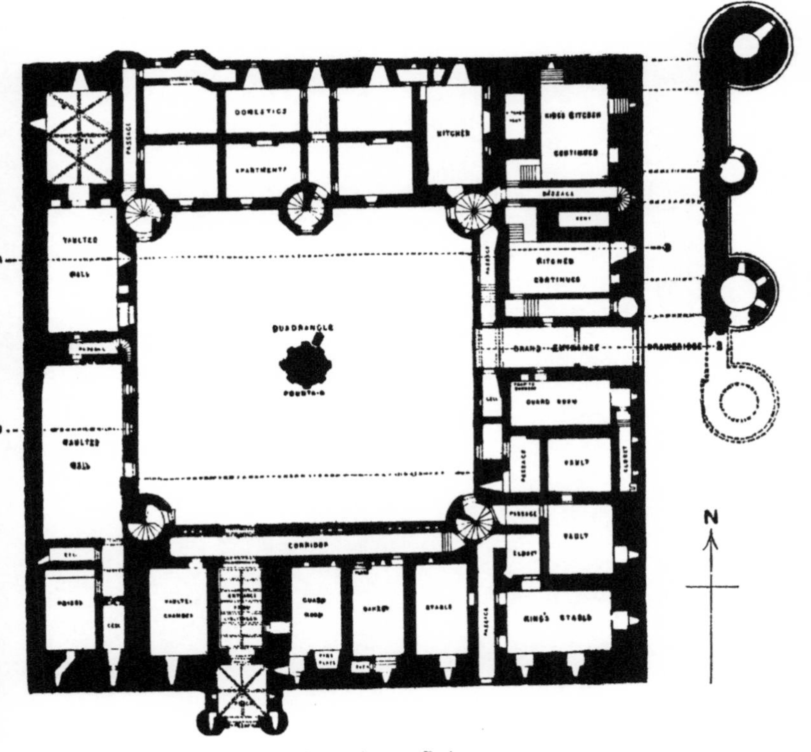

Linlithgow Palace

(*Plan of Ground or Entrance Floor*)

so much enjoyed the sport the neighbourhood provided that a more substantial building of stone was erected. Much of the land in the vicinity was crown property, and it was natural that a manor-house should be built, from which the domain was administered. Taken from the English after Bannockburn, and destroyed by the Scots, the manor-house was restored by Robert II, but was again burned to the ground in 1424.

The researches of Dr Ferguson have proved that the palace was the work of James I. Subsequent monarchs repaired, altered, decorated, and even enlarged it, but without departing from the original design. It was begun in 1424, and during the succeeding eleven years a sum of £4500 was spent. Such a considerable amount must have produced a corresponding result.

The palace was almost square in shape, being 174 feet long by 168 feet broad. The enclosed court is 91 by 88 feet, indicating a divergence from squareness imperceptible to the eye. Exclusive of turret chambers, the palace rises to a height of four storeys. Placed upon a green knoll, between the town and the loch, it commands an extensive view east and west along the valley.

The only portion that can with certainty be ascribed to James I is the west side or back, which has remained practically untouched except by time. Its general appearance from without is that of a wall, the grimness of which is accentuated by the few small windows. It contains the room in which Queen Mary was born, a chapel with a ribbed ceiling, and an oratory, called Queen Margaret's Bower, from a tradition that there the wife of James IV

Room in which Mary Stuart was born, Linlithgow Palace

looked out for his return from Flodden. The remainder of James's building has disappeared, the materials having been used in the extensive alterations of James IV and James V.

James IV carried on building at the palace during most of his reign. The magnificent Lyon Chamber or Parliament Hall on the east side, and the Chapel on the south side were most probably his work.

Since the reign of James V no new work has been done at the palace, and it is to him that we owe not only most of the decoration, but also one or two important additions. Previous to his reign the principal entrance was from the east across a drawbridge. James built a massive gateway on the south or town side. Above the entrance are carvings representing the Collars of St Michael, the Golden Fleece, the Garter, and St Andrew or the Thistle. The last order was instituted by him though it fell temporarily into abeyance during later troublous times. To him are also due the gallery on the south side of the court, and the beautiful fountain, now in ruins, but a copy of which stands in front of Holyrood Palace. Painters, gilders, and glaziers completed elaborate decorations. The windows of the Parliament Hall and the Chapel were filled with stained glass. Well might Mary of Guise exclaim that she had never seen a more princely dwelling.

During the reign of James VI the north wall, which had been propped up by his grandfather, fell down and was rebuilt. This explains why the ciphers of James and his son Charles are on many of the window pediments facing the courtyard.

The final blow came on the 1st of February, 1746, when the palace was burned down by the carelessness of the soldiers of the Duke of Cumberland, not Hawley's dragoons as Scott says. "So with the Stuarts," remarks

South Gateway, Linlithgow Palace

Dr Ferguson, "their chief palace came to an end." For long neglected and even subject to depredations, it was at length taken over by Government, and it is now under H.M. Board of Works, which secures it against further decay. Its association with Queen Mary has given it an interest which is scarcely excelled even by Holyrood. A proposal made by Lord Rosebery in 1910 to restore it as a national memorial to King Edward has not so far met with practical response. Indeed, from one point of view it seems better to leave it as it is—a ruined monument to a fallen cause.

22. Communications—Roads and Railways.

As we have already noted, Linlithgowshire at the present day has no centre of natural importance to act as a focus for roads or railways. It is a county to be passed through on the way from Edinburgh to Glasgow and to the north of Scotland. But before the construction of roads two places were useful as half-way houses—Linlithgow, between Edinburgh and Stirling; and Bathgate, conveniently placed on the road to Lanark. At an earlier date, to go back to Roman times, Cramond was at the end of a road coming from the south, and Carriden, near Bo'ness, was the terminus of the Roman wall and road connecting the Clyde with the Firth of Forth. It is certain also, that Carriden and Cramond were joined. This wall and road marked the extreme limit of effective Roman occupation, and on that account the northern part of the county must have often presented a busy scene.

The Roman road extended from Carriden or rather Bridgeness to Inveravon, where it crossed the river by ford. It had a base of fairly large stones, above them were smaller stones forming in section a rounded surface.

There were no squared kerbs at the sides as was usual with more permanent highways. The width was 17 feet. The distance from Bridgeness to Inveravon is about 4½ miles. About the road to Cramond less is known, and no traces now remain.

Before the twelfth century bridges existed over the Almond at Cramond, and over the Avon near Linlithgow. There must, therefore, have been some tracks, doubtless of a very rude description. Wheeled traffic did not become at all general till well on in the seventeenth century. Goods were carried for short distances on hurdles or wicker sledges, while for longer journeys pack-horses were used.

In the thirteenth century the monks of Newbattle, who had lands in Lanarkshire, complained that owing to the lack of a public road they had difficulty in passing to and fro. They therefore negotiated with the proprietors, who granted access through their estates. The route selected was along the valley past Broxburn and Bathgate; and this is the line of the main road from Edinburgh to Glasgow, which is one of the best highways in Scotland to-day.

On several occasions during the seventeenth century road-acts were passed. They related chiefly to communications within parishes. It was enacted that all tenants and cottars were to contribute as much labour as was required to keep the ways in good condition. The landlords were obliged to assist if necessary. Provision was made later for the commutation of labour by a money payment. In 1633, when preparations were being made to receive Charles I, it was discovered that the roads had been encroached upon, and they were ordered to be widened again.

The first turnpike road in Scotland was that from Edinburgh to Queensferry, opened in 1751. Toll-houses were built at various points, and some of these are still in existence. About the beginning of the nineteenth century Telford and Macadam introduced their methods of road-making, and thereupon a great improvement took place. These roads were maintained by means of tolls, but this system was finally abolished in 1883. At the present day each burgh is responsible for roads and streets within its boundaries, while the County Council looks after the rural communications.

The gradient of the roads in Linlithgowshire depends entirely on the direction in which they run. Those which follow the two remarkable east and west valleys are generally very level, except where transverse streams are met with. The severest hills on an east and west road are that which begins at Cramond Bridge and the famous "Hawes Brae" near Queensferry; but this road is north of the Linlithgow valley. The roads leading north and south cross the ridges of volcanic rock, and present a succession of ups and downs resembling a switchback railway. Owing to the small number of coast towns there is no coast road in the county such as exists between Edinburgh and Aberlady. From Bo'ness a good coast road crosses the Avon to Grangemouth. The main highways are those from Edinburgh to Stirling by Linlithgow, and to Glasgow by Bathgate. The Queensferry road leads to the famous and ancient ferry. Here the opposite shores of the Firth approach within a mile of one another. Traffic is carried by a small steamer, and the passage is used chiefly by motor-cars and tradesmen's vans.

The Union Canal begins at Edinburgh, crosses the Almond by a fine aqueduct half-way between Midcalder and Kirkliston, and traverses the county by means of the midland valley. Another aqueduct carries it over the Avon about a mile above Linlithgow Bridge. Throughout the whole distance from Edinburgh to Camelon, near Falkirk, where it joins the Forth and Clyde canal, no locks are required, which is a sufficient testimony to the level character of the surface. The canal was completed in 1822. The traffic is very small, being confined chiefly to the carriage of coal. Like the Forth and Clyde canal, it is the joint property of the North British and Caledonian Railway Companies, who purchased it to prevent what they considered unfair competition. There is no reason whatever why canals should not convey heavy goods of a non-perishable character which would not suffer by delay, but of this there is no prospect as long as the canals remain in the hands of railway companies.

Two railway companies have lines in the county. To take the less important first, the Caledonian line to Glasgow just cuts through the extreme south for a little more than a mile, with a station at Fauldhouse. Two or three mineral lines branch off at this place and at Addiewell. The Company has also running powers over the North British line past Linlithgow, for trains to Stirling, Perth, and Aberdeen.

Three important North British routes pass through the county. The first crosses the Almond below Kirkliston, and, just beyond Dalmeny station, enters the Forth Bridge. This bridge is the greatest engineering feat of its kind in the world. It is constructed on the cantilever or balance principle, the only one practicable where such enormous spans are required. Had not the island of Inchgarvie been there on which to found the bases of the huge tubes, the enterprise would have been much more difficult if not impossible. Fortunately the land rises quickly from the shore to a considerable height on both sides, so that costly embankments were not necessary. Lofty viaducts, however, lead from each end to the beginning of the cantilever arms. These branch out from three bases, the northern one being on the Fife shore, the middle one on Inchgarvie, and the third on a foundation 80 feet below high water. The total length including the viaducts is 1 mile 972 yards, and the line itself is at a height of 152 feet above high water. To prevent rust the iron is painted; and it is said that the workmen have no sooner reached one end than it is time to begin again at the other.

The second North British route crosses the Almond above Kirkliston, and proceeds by way of Winchburgh and Linlithgow to Falkirk and Glasgow. Long viaducts carry the line across the valleys of the Almond and Avon. From Winchburgh is a line to Dalmeny used chiefly by trains between Glasgow and Aberdeen.

Thirdly, near Ratho, is a junction whence a line leads past Bathgate and Airdrie to Glasgow, by which a rich

Forth Bridge

mineral district is served. Bathgate is a junction of some importance. Southwards runs the Morningside and Coltness branch, while north-westwards a line leads to Blackston Station on the Slamannan Branch, and also to Bo'ness. The latter line is also reached from Manuel, which is just over the border in Stirlingshire. Merely to mention the numerous branches and mineral lines would occupy much space, but enough has been said to show that the county is abundantly served by railways.

There are no tramways in any of the towns, but motor-buses run several times daily from Edinburgh to Queensferry, and to Broxburn and Uphall.

23. Administration and Divisions.

The history of the early local administration of Scotland is extremely obscure, and open to much conjecture, particularly in regard to Lothian, which was successively occupied by different races. On King Edgar's death his younger brother David was made Prince of Lothian and Cumbria. One of his charters is addressed to all his faithful Thegns and Drengs of Lothian and Teviotdale. It proves that by this time the administration of Lothian was Saxon in character. We thus learn that the lands were "held partly by the Thanes in feu-farm and worked by the servile class," and partly by "kindlie tenants and free farmers." The thane was responsible to the Prince or King for the good order of his thanage.

Gradually with the influx from the south, this state of matters passed away, and was merged in the Norman or feudal system, which attained a higher development in Scotland than in England. Even in the reign of David I the Exchequer Rolls mention "the King's Lordship of Linlithgowshire." The royal lands, however, soon passed into the hands of favourite subjects on a feudal tenure or were gifted to religious houses. At the present day the

Town Hall, Linlithgow

peel or park of Linlithgow is all that remains as crown or national property.

In the reign of Robert II, West Lothian was reduced from the status of county to that of a constabulary under the Sheriff of Edinburgh. It remained in this subordinate position until the reign of James III, who restored it to the rank it now occupies.

For judicial and administrative business, a sheriff-principal presides over the three Lothians. Linlithgowshire has a sheriff-substitute resident in the county town. It is interesting to note that till the sixteenth century Bathgate and the surrounding territory formed part of the sheriffdom of Renfrew. This was because that district was a portion of the dowry of Marjory, Robert the Bruce's daughter, on her marriage with Walter the sixth Steward, baron of Renfrew.

Linlithgowshire contains eleven entire civil parishes and part of another, and eleven ecclesiastical parishes with parts of five others. There are two royal burghs, Linlithgow and Queensferry; and four police burghs, Armadale, Bathgate, Bo'ness and Whitburn. The county is divided into twenty-four electoral divisions, which return twenty-four members. Bathgate and Linlithgow have district Committees of Management. The Lord-Lieutenant is the Earl of Rosebery, and there are three Deputy-Lieutenants. The powers of the County Council are many and varied, ranging from control of technical education to maintenance of the roads.

Parochial matters are managed by elected Parish Councils. Their chief duty is the administration of the poor laws, which include the provision and upkeep of workhouses. Part of the rate they are empowered to levy is spent by the School Boards, elected bodies responsible for education within their areas.

The county constabulary is partly under the County Council and partly under the Commissioners of Supply. The force consists of one superintendent, four inspectors, eight sergeants, and 48 constables. The headquarters are at Linlithgow, and there are two divisions; Linlithgow with sections for Bo'ness and Broxburn, and Bathgate with sections for Armadale and Whitburn. The royal burghs control their own police.

For purposes of mines inspection the county is in No. 1 (Scotland) district, and in No. 46 (Edinburgh) district for factory inspection.

The military force is in the Scottish Command with headquarters at Edinburgh. With the exception of a lieutenant, a sergeant and 25 men in Blackness Castle, the army in the county is represented by the Territorial Branch, which consists of the 10th (Cyclist) Battalion Royal Scots, forming part of the Lothians Infantry Brigade. The Territorial Association established by Act of Parliament in 1907 has for its President, the Lord-Lieutenant. There are eight military members, two representatives elected by the County Council, five co-opted members, and a Secretary. The battalion is composed of eight companies stationed throughout the county with headquarters at Linlithgow. The value of cyclists as scouts has passed beyond the stage of requiring demonstration, and is now an accepted fact.

The burghs of Linlithgow, Hamilton, Lanark, Airdrie and Falkirk, form a combination known as the Falkirk Burghs and return one representative to Parliament. Queensferry joins with Stirling, Culross, Inverkeithing, and Dunfermline (Stirling Burghs) in returning another. The remainder of the shire, with an electorate in 1911 of 11,480, has also a representative.

24. The Roll of Honour.

Royal residents in Linlithgow, and royal visitors, have already been mentioned. Here we recall the memory of Mary Stuart alone, who was born in the Palace, 8th December, 1542. For that, if for no other reason, the town is dear to the people of Scotland. During her troubled reign she came several times to Linlithgow. One of her faithful attendants, Mary Seaton, was one of the "loyal Seatons" of Niddry Castle.

Of noble families, the barony of Livingston, now extinct, was founded by David II. In 1599 Alexander, eldest son of the sixth baron, was created Earl of Linlithgow and made keeper of the palace, an office held by his successors in the earldom till 1715. From their ownership of Kinneil House and estate the Hamiltons are closely connected with the county. Douglas, the eighth duke, whose mother was the noted beauty Elizabeth Gunning, was keeper of the palace from 1770 to 1790.

The Hopetoun family is one which has shed lustre on the county. Its founder was Thomas Hope, Lord Advocate in the reign of Charles I. He made his name by his defence of the six recalcitrant ministers who were tried and condemned by a packed jury at Linlithgow in 1606. His great-grandson, Henry Hope, was created Earl of Hopetoun in 1703. His descendant, the late Marquis of Linlithgow, who died in 1908, was Governor-General of Australia when the Commonwealth was formed in 1901.

The Dundases of Dundas and the Dalyells of Binns, though not noble, are very ancient families. General

Mary Stuart

Thomas Dalyell raised a regiment, the Scots Greys, which he employed with so much effect against the

Covenanters. Another soldier was Colonel James Gardiner who was born at Burnfoot, east of Bo'ness. He appears in *Waverley*, where his death at the battle of Prestonpans is strikingly related. Two greater contrasts than Gardiner and Dalyell of Binns can scarcely be imagined.

Churchmen and statesmen are not prominent in the records of the county. Naturally when the court was in residence the great officials of the realm were in attendance, but that connection is somewhat accidental. We have already mentioned Bishop Wilfrid of York, who founded a monastery at Abercorn. Another bishop, David de Bernham of St Andrews, consecrated St Michael's Church at Linlithgow in 1242.

Under James V, Patrick Hamilton of Kingscavil and Henry Forest, a Benedictine friar, both natives of Linlithgow, were burnt at the stake, the earliest Scottish martyrs for the reformed doctrines.

In the chancel of Uphall Parish Church are buried the two brothers, the Hon. Harry Erskine, and Thomas, Baron Erskine—the former famous at the Scottish Bar; the latter, one of the greatest forensic orators, and Lord Chancellor of England.

Of poets and authors generally, not a few have had some connection with the county. Sir Robert Sibbald wrote a history of the county, and lived at Kipps House in the seventeenth century. William Hamilton of Bangour, who died in 1754, was the author of the immortal ballad, "The Braes of Yarrow,"

> "Busk ye, busk ye, my bonny, bonny bride."

After taking part in the '45 he escaped to France, but returned home, and succeeded to his estates. He died, however, at Lyons. The Rev. William Wilkie, D.D., a native of Echline, wrote the *Epigoniad* in nine books, a poem founded on the story of the sack of Thebes. In his day he was called the Scottish Homer, but his work is now as dead as many another epic. During the nineteenth century four men of letters are worthy of being recorded. The first is John Campbell Shairp, born at Houston House, near Uphall, who became Principal of St Andrews University, and Professor of Poetry at Oxford. He wrote many works of criticism, and contributed to *belles lettres* generally, but he will go down to posterity as the author of the lovely ballad—"The Bush aboon Traquair." The second is Alexander Smith, at one time librarian of Edinburgh University, whose association with Linlithgow is residential. During the sixties his poems had great vogue, but, partly because some ill-advised persons set him up as a rival to Tennyson, his poetry fell into ridicule as belonging to the "Spasmodic School." His prose has a fine literary quality, the best being *A Summer in Skye*, and *Dreamthorp*. The latter is a collection of essays, written chiefly at Linlithgow, and pervaded by a subtle flavour of the town, its scenery and memories. Third is Dr John Brown, who was born at Whitburn in 1784. He is renowned as a theological writer with a special gift for exposition. Dugald Stewart, the distinguished Professor of Moral Philosophy in the University of Edinburgh, resided in Kinneil House for the last twenty years of his life. During this period

Sir James Young Simpson

he published most of his best known works on philosophy, which are now regarded as explanatory of and in a less degree supplementary to those of his master, Reid of Aberdeen.

Linlithgowshire has given birth to several distinguished men of science, inventors, and engineers. Sir Wyville Thomson, born at Bonsyde, was the leader of the "Challenger" Expedition, which did so much to advance the science of Oceanography. Henry Bell, to whom is due the chief, if not the sole, credit of inventing the application of steam to vessels, was born at Torphichen in 1767. The "Comet" launched at Helensburgh in 1812 was the first steamer successfully to accomplish a voyage at sea. Dr John Roebuck of Sheffield, who created the iron industry at Carron, resided at Kinneil House, and was buried in the new churchyard at Carriden.

Last, but perhaps greatest of all, is the discoverer of chloroform as an anaesthetic in surgical operations. Sir James Young Simpson was born at Bathgate in 1811. He became Professor of Midwifery at Edinburgh University in 1840, was knighted in 1854, and made a baronet in 1866. For years he had been experimenting with a view to discovering some means of alleviating the pain of patients under the knife. One day it is said that he received a letter from Dr James Waldie, a Linlithgow practitioner, advising him to try chloroform. It was done, and found practicable. This is one of the greatest discoveries that has ever been made in medical science. Well might Simpson take for his motto the beautiful words, "victo dolore." Robert Liston, another eminent surgeon, was born in Ecclesmachan Manse in 1794.

25. THE CHIEF TOWNS AND VILLAGES OF LINLITHGOWSHIRE.

(The figures in brackets after each name give the population in 1911, and those at the end of each section are references to pages in the text.)

Abercorn. The parish of Abercorn (933) is the oldest in the county, and one of the oldest in Scotland. It contains no towns, but has several villages and hamlets. Abercorn is British, and means "mouth of the Cornar or Curnig," i.e. horned stream. (pp. 39, 43, 56, 79, 80, 97, 106, 127.)

Armadale (4739), a town and police-burgh, is in the civil parish of Bathgate, but for ecclesiastical purposes is a *quoad sacra* parish. In the neighbourhood are coalfields, ironstone quarries, and deposits of clay. The industries of the town are steel-works, brick, tile, and fire-brick yards. The population is growing rapidly. Near Armadale is the Southern District Combination Hospital, erected in 1901. (pp. 68, 72, 105, 123, 124.)

Bathgate (8226), a market-town and police-burgh, was created a burgh of barony by Charles II in 1661, and a free and independent burgh of barony in 1824. Near the town are coal and iron mines. The industries include foundries, spade and shovel works, and a distillery. No less than seven fairs are held annually. Bathgate Academy for secondary education stands in a good position on the outskirts. Its erection was due to the munificence of John Newland, a native of the burgh, who made a fortune in Kingston, Jamaica. It is managed by a body of trustees, and is a commodious and well-arranged building. The old parish church, erected in 1737, was demolished in 1882. On the same site was erected the present church, a fine building in the Gothic style. Bridgend is a western suburb. Woodend and Durhamtoun are colliery villages in Bathgate parish. Bathgate is

Bathgate Academy

a railway junction of some importance, and the centre of a busy mining district of rapidly growing population. The origin of the name is obscure, the earliest spellings being Bath chet, Bath ket and Bat ket. (pp. 6, 9, 11, 27, 42, 51, 61, 63, 68, 70, 74, 77, 88, 89, 90, 97, 101, 105, 116, 117, 119, 121, 123, 124, 130.)

Borrowstouness, generally written, and always pronounced **Bo'ness** (10,866), is the largest town, and the only considerable seaport in the county. It is composed of the combined burghs of Bo'ness and Carriden, which are in separate parishes. The parish of Carriden has doubled its population in 20 years. There are three large collieries which produce the best coal in the county. Excellent freestone and whinstone are quarried, the former for building purposes, the latter for road-metal. A large distillery produces chiefly yeast and other by-products, chemical works manufacture manure, saw-mills turn out pit-props, from cabinet-making works come door and window frames. Two potteries have a large output of earthenware and china, from dinner services to mantelpiece ornaments. The total value of the exports and imports is over £600,000. Ample provision is made for elementary education, while Bo'ness Academy, also under the School Board, is for more advanced pupils. The name of the town is English, meaning, "the burgh town on the naze or cape." From its situation on a promontory it would naturally attract sea-faring people.

To the south, just across the Roman Wall, is Newtoun, a mining village. Grangepans, and Bridgeness are eastern suburbs. The former points to the existence of salt-works. (pp. 4, 5, 8, 12, 23, 27, 29, 30, 33, 34, 41, 42, 43, 44, 45, 51, 54, 59, 63, 64, 65, 66, 68, 70, 72, 74, 75, 76, 77, 78, 88, 92, 104, 105, 109, 110, 115, 117, 121, 123, 124, 127.)

Broxburn (9000), Gaelic, the burn of the brock or badger, a growing town which may be regarded as the metropolis of the shale-oil industry. It is situated on the Brox Burn, a small stream which joins the Almond at Newliston. The Union Canal passes through the middle of the burgh. Shale-mining and oil-distilling are the occupations of the majority of the inhabitants. The Broxburn Oil Company has works in the town; but Pumpherston, Hopetoun, and other mines are close by. It is here that the great "bings," or mounds of spent shale, are most numerous and prominent. The town sits astride of the main road from Edinburgh to Bathgate and Glasgow, but the railways are some way off, the nearest station being Drumshoreland, one mile distant. Motor omnibuses ply between Edinburgh, Broxburn, and Uphall. (pp. 11, 26, 27, 39, 59, 70, 105, 109, 116, 121, 124.)

Dalmeny (4442) is a parish in the north-east. The village of the same name lies behind the crest of the bluff above Queensferry. Shale is worked, and a new pit was opened recently. During the thirteenth century the manor of Dalmeny was in the hands of the Moubray family. Later, through forfeiture, it passed to the Menteiths. It was acquired in the reign of Charles II by Archibald Primrose, who was created Earl of Rosebery and Lord Dalmenie in 1703. The latter title is now used by the eldest son of the Earl of Rosebery. (pp. 9, 13, 15, 23, 27, 29, 30, 34, 35, 36, 39, 44, 54, 58, 59, 61, 70, 96, 106, 109, 119.)

Ecclesmachan (parish, 1449) is a village about two miles north-west of Broxburn, on the ridge separating the two longitudinal valleys. It is surrounded by shale-mines, while freestone and sandstone are worked in the neighbourhood. The Niddry Burn flows through the village on its way to the Almond. Binny Craig with its conspicuous precipice lies to the west. The family of Binny or Binning traces its descent from William Binnock, who according to tradition took the Peel of Linlithgow from the English in 1313. The story goes that he concealed eight men in a cart of hay which he stopped in the entrance and so prevented the descent of the portcullis. It is the kind of tale that one likes to believe, but it rests on no secure foundation. In 1250 the place-name is spelt Eglismanin, the church of Machan or Manchan, a saint who probably flourished in the latter half of the sixth century. Near Ecclesmachan is one of the few mineral wells of the county—Bullion Well. It is said to resemble the Moffat water, but as yet there is no attempt to create a spa. (pp. 74, 130.)

Fauldhouse (3923) is a flourishing village in the extreme south, close to the boundary. It is in a bleak district, surrounded by moors, which are dotted with the ungainly pit-head machinery of the coal and iron mines. It is the highest village in the county (751 feet). (p. 118.)

Kingscavil, two miles east of Linlithgow on the Edinburgh road, is on the western edge of the oil-shale region. The name means king's allotment or share of land. (pp. 85, 87, 127.)

Kirkliston (population of Linlithgowshire part of parish, 4467) is a village on the left bank of the Almond, and therefore close to the eastern boundary. Part of the parish of Kirkliston steps over the Almond into Midlothian. In the village is a distillery producing malt extract. Several shale-mines are in the vicinity. The fine old parish church is in the Norman style. Here is the burial ground of the Dalrymple family. One who lies there is Elizabeth Dundas, wife of John, first Earl of Stair, and the prototype of Lucy Ashton in the *Bride of Lammermoor*. The mansion-house of the Stair family is at Newliston, close by. The plantations of trees in the park are said to be arranged as a plan of the battle of Dettingen, where the second Earl distinguished himself. (pp. 7, 34, 39, 43, 56, 63, 90, 97, 98, 118, 119.)

Linlithgow (4003), the county town, is an ancient royal burgh. The oldest part of the town extends from the steep ascent to the palace westwards along the south shore of the loch. It now consists mainly of one mile-long street, broad at both ends and very narrow in the middle, with many closes, wynds, or vennels darting off at right angles from either side. At the eastern end of the narrowest part is a fine square, the north side of which is occupied by the Town Hall. In front of the Hall is the Cross Well, restored in recent times, the original of which was built in 1520. The industries of the town include paper-making, glue-making, tanning, distilling, and the manufacture of safety fuses. Despite these varied employments, the population (unlike that of the other parishes and towns in the county) is declining in numbers. The arms of the town bear the figure of St Michael, its patron saint, and a black dog tied to a tree, the latter probably in allusion to the hunting pursuits of the Scottish kings. The motto is, "My fruit is fidelity to God and the King." Three miles to the east is the village of Bridgend about a mile from Kingscavil, and like it surrounded by shale-mines. (pp. 2, 3, 4, 5, 6, 9, 10, 11, 15, 16, 17, 18, 25, 26, 28, 29, 30, 32, 33, 34, 40, 41, 43, 49, 50, 51, 53, 61, 63, 64, 67, 68, 73, 74, 75, 79, 80, 81, 82, 83, 84, 85, 86, 87, 88, 89, 99, 100, 104, 105, 106, 109, 110, 115, 116, 117, 118, 119, 122, 123, 124, 125, 127, 128, 130.)

Livingston (3714), a parish east of Bathgate. The village of the same name is on the left bank of the Almond. Several shale-pits and oil-works are in the neighbourhood. (pp. 7, 13, 31, 125.)

Newton is a small agricultural village in the parish of Abercorn.

Niddry, a hamlet in the same parish, gives the title of Baron to the Hope family. The chief employment is shale-mining. Niddry is a British place-name. (pp. 70, 85, 104, 125.)

Philipstoun, in the parish of Abercorn, depends entirely upon the shale-mines and oil-works of Messrs James Ross and Company. It is on a by-road, but has a station on the main Glasgow line of the North British Railway. (pp. 27, 40, 59, 70.)

Queensferry or South Queensferry (2812), on the Firth of Forth, where it narrows to a little over a mile, has no industries of importance except the shale-mines, a short distance inland. It is a royal burgh dating from the reign of Charles I. About half a mile to the west is the Forth Bridge. Below the bridge is the chief anchorage for the Fleet. Its coming brings great, if temporary, prosperity to the otherwise sleepy little town. Some old buildings are clustered round the little harbour, but few traces remain of the nunneries and monastery which existed before the Reformation. Port Edgar, half a mile to the west, is associated with one of the brothers of Queen Margaret. (pp. 8, 33, 34, 37, 38, 39, 45, 74, 75, 99, 106, 117, 121, 123, 124.)

High Street, Queensferry

Torphichen (4046) is a parish on the western boundary south of Linlithgow, and spreads over both sides of the volcanic ridge, the eastern end of which is at Binny. Another ridge stretches from near Bathgate northwards, thus forming a T. The parish is mainly agricultural and pastoral, but the southern part has coal and iron mines. In the south-west is Blawhorn Moss, one of the few uncultivated areas in the county. The hamlet itself is interesting for its remains of a hospital or preceptory of the Knights of St John of Jerusalem. In the old mill was born Henry Bell, the creator of the "Comet" steamboat. The village is situated at a height of 600 feet above the sea, which makes it one of the highest in the county. Westfield paper-mill is about two miles away and close to the Avon, but it belongs to Bathgate rather than to Torphichen as the means of communication favour the former place. The name Torphichen is Gaelic, and means "raven hillock." (pp. 11, 15, 17, 27, 53, 68, 89, 98, 105, 130.)

Uphall (parish, 12,767) is on the Glasgow road, one mile west of Broxburn. Situated in the heart of the shale-mining district, it has also sandstone quarries, coal-mines, and deposits of ironstone. There is a factory for bone manure. The famous Pumpherston oil-works are between Uphall and Midcalder. Two miles to the west but in the parish of Ecclesmachan is Bangour (Gaelic, Goats' Stead) Asylum. It is under the management of Edinburgh Parish Council, and is occupied by feeble-minded poor from the capital. The establishment is on the most modern lines, resembling a miniature "garden city," and it is considered to be in every respect a model of what such a refuge should be. (pp. 51, 59, 70, 71, 97, 110, 121, 127, 128.)

Whitburn (1875), East Whitburn, East Benhar and Stoneybury Colliery form a group of mining villages in the south of the county. (pp. 13, 51, 123, 124, 128.)

Winchburgh is 10½ miles from Edinburgh on the Linlithgow road. Here the famous Oakbank Company erected extensive oil-works in 1901. There are also brick and tile works. The Union Canal passes beside the village. Winchburgh is an English name. (pp. 74, 109, 119.)

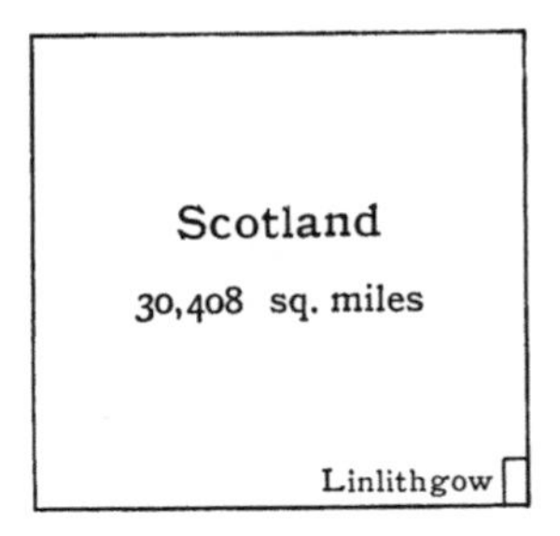

Fig. 1. Comparative areas of Linlithgowshire (121 sq. miles) and all Scotland

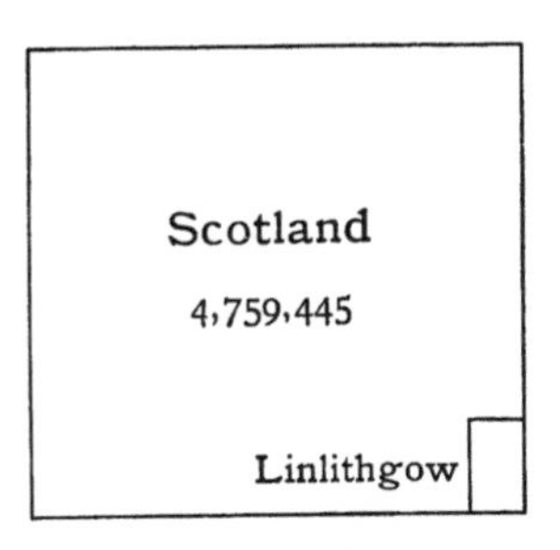

Fig. 2. Comparison in Population of Linlithgowshire (79,456) and all Scotland in 1911

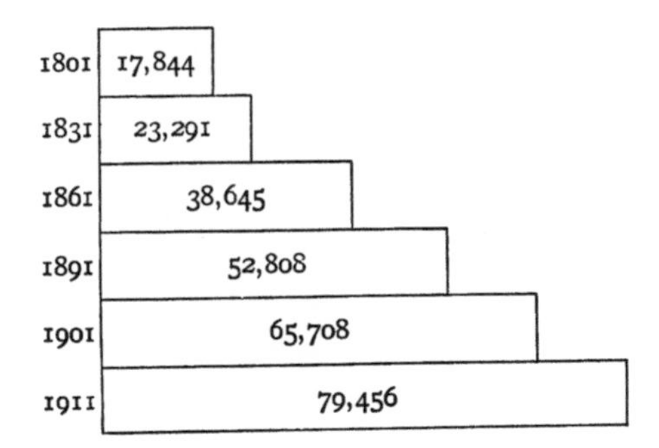

Fig. 3. Growth of Population in Linlithgowshire

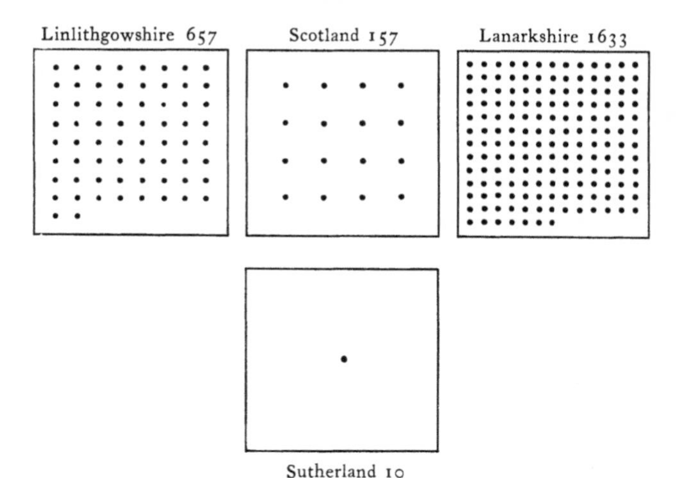

Fig. 4. Comparative density of Population to the square mile in 1911

(Each dot represents ten persons)

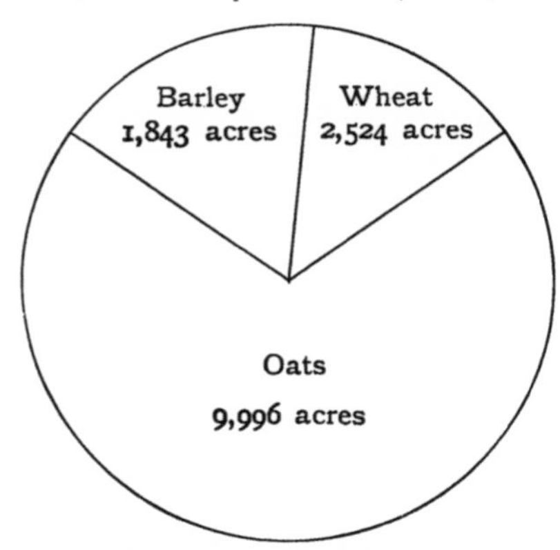

Fig. 5. Comparative areas under Cereals in Linlithgowshire in 1911

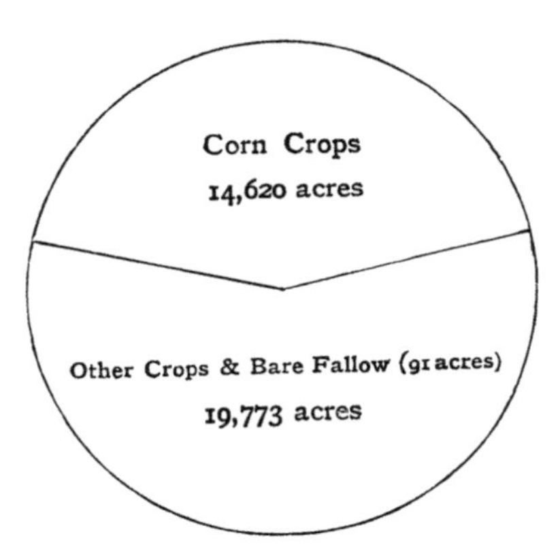

Fig. 6. Proportionate area under Corn Crops compared with that of other land in Linlithgowshire in 1911

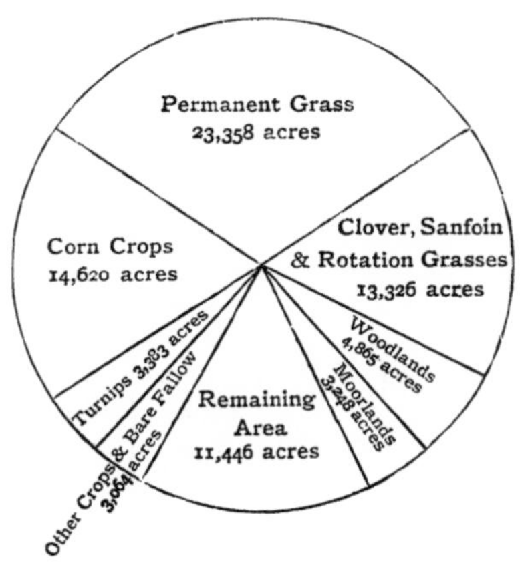

Fig. 7. Comparative areas of land in Linlithgowshire in 1911

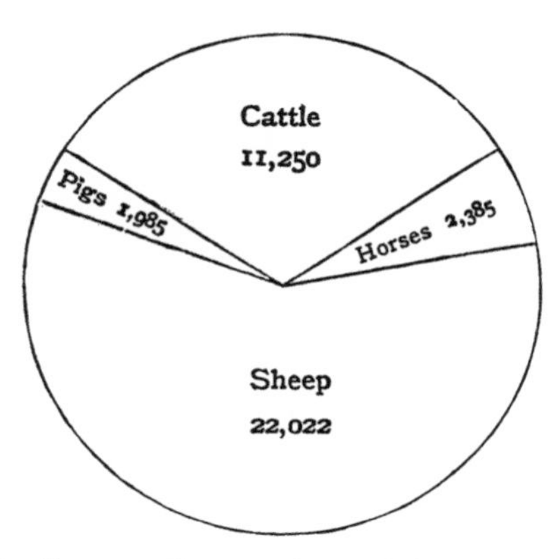

Fig. 8. Comparative numbers of different kinds of Live Stock in Linlithgowshire in 1911

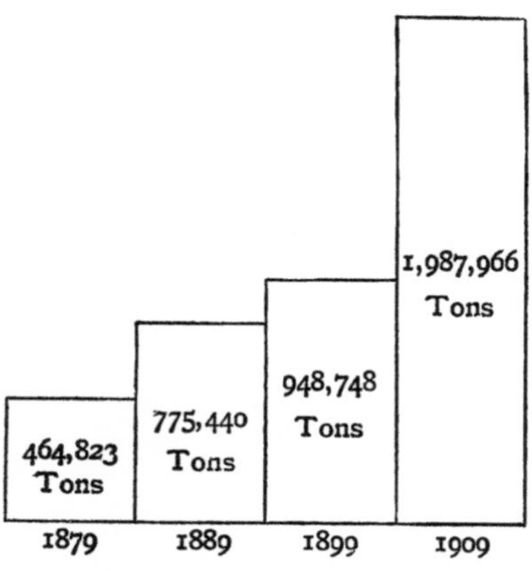

Fig. 9. Progressive output of coal in Linlithgowshire

www.ingramcontent.com/pod-product-compliance
Ingram Content Group UK Ltd.
Pitfield, Milton Keynes, MK11 3LW, UK
UKHW042145280225
455719UK00001B/121